THE
BEGINNER'S GUIDE TO
FLIGHT
INSTRUCTION

BY JOHN L. NELSON

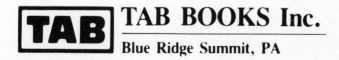

TAB BOOKS Inc.

Blue Ridge Summit, PA

FIRST EDITION

FIFTH PRINTING

Printed in the United States of America

Reproduction or publication of the content in any manner, without express permission of the publisher, is prohibited. No liability is assumed with respect to the use of the information herein.

Copyright © 1983 by TAB BOOKS Inc.

Library of Congress Cataloging in Publication Data

Nelson, John Lewis, 1926-
  The beginner's guide to flight instruction.

  Includes index.
  1. Flight training.   2. Private flying.   I. Title.
TL712.N44   1983   629.132′5217   82-19363
ISBN 0-8306-2324-8 (pbk.)

Questions regarding the content of this book
should be addressed to:

  Reader Inquiry Branch
  TAB BOOKS Inc.
  Blue Ridge Summit, PA 17294-0214

Contents

Preface

This text is directed to the subject of reducing the cost of obtaining a private or commercial certificate through understanding the true requirements of a certificate. In this day of inflation and rapidly changing interest rates, both flying and flight instruction have become expensive ventures. While flying will never be a truly low-cost activity, you can certainly minimize the dollars spent and enjoy your time in the air to the very utmost. There is much you can do to obtain a maximum of value from your "flying" dollars, provided that you will extend the effort. Henry George expressed it well when he wrote in *The Science of Political Economy*, "The value of a thing is the amount of laboring or work that its possession will save the processor." By understanding how we learn and what we must learn, by carefully selecting our flight training program, and by appling ourselves diligently, we can and do achieve the value that makes flying worth the dollars spent.

A note regarding the various cost figures contained in this text for flight instruction, aircraft operating expenses, and the like. This book reflects cost data in 1981. Since costs tend to follow the inflation rate, it is suggested that such be scaled proportionately to the time frame of the reader.

To maintain the utmost in text accuracy, relevant paragraphs of the FAA *Airman's Information Manual, Aviation Instructor's Handbook*, and other related literature have, at times, been paraphrased and included either in part or in whole. In particular, the private pilot curriculum and glossary of common aviation terms included at the end of this book are selected from FAA training material.

I wish to thank the many people who made this book possible, especially Barbara Nelson for all of the art work and my wife Dolores for manuscript preparation. Many thanks to Joe Christy for editing the text, and to the following, my sincere appreciation for supplying reference material and equipment photographs.

Cessna Aircraft Company.
Beech Aircraft Corporation.
Emery School of Aviation.
Embry-Riddle Aeronautical University.
Aircraft Owners and Pilots Association.
Federal Aviation Administration.
National Transportation Safety Board.

Chapter 1
Getting Started

Gone Flying! In pursuit of an aviation career, or perhaps an adventure, or simply for the enjoyment involved, "Gone Flying" means departure from flat earth and entry into the world of three dimensions. "Gone Flying" signifies separation from the world of mediocrity with its uniform conformity into a domain of freedom and precision, a place where you are master of your fate. No longer is there time to reflect upon stultifying committee decisions, company standard practices, operational procedures, rules, rules, rules . . . it's time now to act wholly as an individual, as pilot-in-command. Whether "Gone Flying" means your first lesson or your 20,000th hour as a senior airplane pilot, the sensation is remarkably similar—a departure from everyday ho-hum into a special world reserved for the achievers of life. Left behind are the *can't dos* or the *won't dos*, ahead we share the airspace with only the *can do's* of life as associates.

Perhaps the sensation that separates flying from the routine of everyday living is the stark realism of each and every decision to be made by a pilot. The process is totally engaging, and wholly absorbing. Every weather decision, fuel computation, or route change is singularly that of the pilot-in-command. Each decision is tested by the elements and tried under the laws of physics—universal and fair, but not without penalty to the careless or reckless. "Gone Flying" is for those who would romance the sky in her many moods, for those who understand the night and feel confident with only the

1

cockpit lights and instruments for guidance. "Gone Flying" is for those who dare.

Approximately 100,000 people per year make the decision to try flying, to see what aviation is all about. For some, it's a matter of career and devotion; for others, business or perhaps recreation, and in many cases, it's a matter of curiosity. Regarding the latter, there is nothing at all wrong with satisfying one's own curiosity and then perhaps going on to other things. Many pilot candidates simply want to experience the sensations of flight or perhaps achieve a goal in life without any intent to participate as a regular activity.

PICKING YOUR FLIGHT TRAINING

Whatever the case, many training routes are available to the candidate. The remainder of this chapter will be devoted to a brief overview of training and the various tradeoffs involved. Chapter 4 is devoted to a detailed discussion of flight training for the private and commercial pilot. Greater detail still is contained in Chapters 8 and 9 which present a complete curricula for the Private and the Instrument/Commercial pilot.

Professional Flying as an Occupation

A first element to consider in the choice of a flight school is your long-term objective in aviation. If your intention is to make flying your career in life, by all means investigate colleges that provide aviation training as a part of a degree program. The majority of the larger colleges offer aviation options and many offer complete programs devoted to aeronautical degrees.

Being realistic about aviation as an occupation, the college degree is rapidly becoming a must. Aviation attracts aggressive people and aggressive people compete with one another for the rewards. A college degree is one of the major tools required to compete successfully today. Allied with this is the fact that our society is becoming technically based at a rate greater than ever before. Even now, complete aircraft are being designed and "flown" in computers without a single piece of aluminum being cut for the actual aircraft. In one instance, computer programming has reached a level of sophistication wherein the computer will "draw" the allowable shape(s) of an aircraft given only the required flight performance parameters (speed, range, ceiling, etc.)! Technical progress cannot be stopped. The next few years will see whole new approaches to aircraft design come into being, due for the most part to inevitable increases in the cost of fuel. A college degree plus your pilot certificate will be absolutely necessary to participate as a professional in the field of aviation in the future.

In the event you are seeking a college wholly devoted to aviation, consider Embry-Riddle Aeronautical University of Daytona Beach, Florida, or Prescott, Arizona. Embry-Riddle offers two and four-year Bachelor of Science degree programs and also Master's degree programs. Study fields include aviation management, maintenance, avionics, computer technology, engineering, and flight careers. Based on 1980 costs, expect to pay approximately $10,500 for a basic four-year degree program. If flight training is to be included, expect to add between $12,000 and $22,000 to the cost of the basic classroom program. (Room and board are not included in the preceding figures.)

Let us take another case: You already have a college degree and simply want to attend a dedicated aviation school to complete pilot training as a full-time student. Or conversely, you wish to complete pilot training as a full-time student first, then pursue other training at a later date at another location. In either event, an institution such as Emery School of Aviation at Greeley, Colorado may fill the bill.* Emery offers a complete program of pilot training for all types of pilot certificates. The school is certified by the FAA and the State of Colorado and is approved for veteran's training. Completion time for a fixed-wing training program through the flight instructor rating is typically 30 to 32 weeks at seven hours of school per day and five days per week. In terms of 1980 dollars costs are typically as follows:

Fixed-Wing Flight Training:

Private Pilot Certi- fication Course	$ 1,802.00
Commercial Pilot Certification Course	8,036.00
Flight Instructor Course (Airplane Single Engine)	1,509.00
Flight Instructor Course (Instrument- Airplane)	1,262.00

---

*It should be noted that Emery School of Aviation does not require a college degree as a prerequisite; a high school education or equivalent is adequate to qualify for pilot training.

Multi-Engine Rating Course	1,227.00
Flight Instructor Course (Multi-Engine)	2,414.00
Helicopter Flight Training:	
Commercial Pilot Certification Course (Helicopter)	9,838.00
Flight Instructor - Helicopter Course	3,152.00
Optional Helicopter Flight Training	
External: Load (Class A & B)	$ 2,486.00
Mountain Flying	3,530.00
Instrument Rating - Helicopter (or)	7,022.00
Additional Instrument Rating (Helicopter)	2,969.00
Instrument Flight Instructor - Helicopter	3,959.00

Fixed Base Operators

The natural choice for business or recreational flight training is the Fixed Base Operator (FBO). Flight training is generally a fundamental business element for most FBO's. As an activity, an FBO may offer a thoroughly professional program or he may offer a marginally acceptable course of flight training. The quality and course content of flight training varies significantly between fixed base operators. Without going into detail, please understand the fact that flight training is generally one of the *least* profitable parts of the FBO business structure. The motivation to offer first-rate flight instruction is therefore subject to the way in which the individual FBO sees flight training as a business element.

Very simply put, *you* must assess candidate FBO's and determine the quality of their program *before* signing on the dotted line. As a guide to evaluating FBO flight training programs at today's prices, the following minimum capabilities should be met:

1. *FAA/VA approval:* Other than single-person operations in remote areas, an FBO should have FAA/VA course approval for his flight training program. The fact that an FBO may "meet" FAA requirements is not sufficient. It's *formal* FAA/VA approval that counts.

2. *Curriculum:* The FBO should review their curriculum in detail with you during your inquiry. Furthermore, a copy should be supplied you with your first flight lesson.

3. *Cost estimate:* The FBO should supply a detailed *written* cost estimate based on an average pilot taking the course. Do not accept verbal-only cost estimates.

4. *Training facilities:* The cleanliness of the classroom(s) is a tip-off to the care and precision with which a course is taught. The same is true with respect to the condition of the training aircraft.

5. *Training aircraft:* In addition to being late models, aircraft should contain at least the following equipment:

☐ Private Pilot: Full instrument panel plus transponder and nav/comm radio.

☐ Commercial/Instrument Pilot: Full instrument panel, plus transponder, dual nav/comm radios, ADF, and DME. In addition, a complex aircraft must be available for final training phases.

6. *Instructor:* A regular instructor should be available for your training program. Constantly changing instructors leads to program inefficiencies in that each new instructor must spend time to assess the student's training progress to date. This is particularly important as your training program enters its final phases. The instructor who has coached you through your program should, by all means, be the instructor that gives you your flight test sign-off.

7. *Training costs:* In this day of rapidly-changing prices, it is difficult at best to accurately describe the cost of flight training. The following represents a typical case for a high quality FBO in terms of 1981 dollars:

Private Pilot Certificate:		
Aircraft rental	53 hrs @	$ 30.00/hr
Flight instructor	33 hrs @	$ 20.00/hr
Ground school		$140.00
Books and equipment		$ 75.00
Flight test		$ 50.00
Total:		$2,515.00

Instrument rating:

Aircraft rental	30 hrs @ $ 30.00/hr
Complex aircraft rental	15 hrs @ $ 65.00/hr
Flight instructor	45 hrs @ $ 25.00/hr
Flight simulator	15 hrs @ $ 40.00/hr
Gound school	$160.00
Books and equipment	$ 50.00
Flight test fee	$ 75.00
Total	$3,885.00

Commercial Rating:

Given the fact that you are already instrumented rated, expect to spend between $1000 and $2000 to obtain your commercial certificate, depending upon your experience level.

For those desiring advanced ratings, the following are typical 1981 costs:

Flight instructor	(single-engine, land)	$1700.00
Flight instructor	(instrument)	$1300.00
Flight instructor	(multi-engine, land)	$3700.00
Multi-engine rating		$2800.00
Airline transport pilot	(single-engine, land)	$2500.00

Freelance Flight Instructors

Another means of obtaining flight instruction is through a freelance flight instructor. There are many fine freelancers, who do an excellent job of providing flight instruction. For some, freelancing is a sole means of support; for others, it is an avocation of serious interest. Many freelance flight instructors have excellent aviation backgrounds, particularly retired FAA examiners, ex-USAF pilot instructors, and ex-air carrier/air taxi pilots. The freelancer generally teaches because he enjoys flying; as a result, his rates are often lower than FBO rates in that he has no overhead costs to support.

In general, a freelance flight instructor will have access to a training aircraft, often through a club or perhaps a private agreement with an aircraft owner. Once again, the reduction in overhead can provide an aircraft rental cost reduction to the student. Typical freelance flight training prices for 1981 are:

Flight instruction	$15.00/hr
Instrument flight instruction	$20.00/hr
Ground school instruction	$5.00—$7.00/hr
Aircraft rental	$20.00—$25.00/hr
Complex aircraft rental	$50.00/hr

Looking at an overall flight training program, a reduction in cost of 25% to 30% may be possible if accomplished with a freelance instructor rather than an FBO. Then why not take this route? Indeed, why not, if the quality of the training is the same? And *there* you have the question—just how do you assess a freelance course of training?

Before you is the same task as with the FBO—determining the quality of a program of instruction before signing up. The procedure is similar, only the questions are different. As a minimum, the following should have positive answers:

1. *Flight instructor qualifications:* Determine that the flight instructor is qualified to give instruction for the rating you seek. For example, only a Certified Flight Instructor-Instrument (CFII) can provide instrument instruction; a regular CFI (Certified Flight Instructor) cannot. Remember too that all flight instructors must renew their certificate every two years. Will your freelance instructor's certificate expire partway through your training? Does he or she intend to renew?

2. *Curriculum:* FBO or freelance, a flight training curriculum should be employed and a copy made available to the student pilot.

3. *Cost estimate:* Just as in the case of an FBO, the freelance flight instructor should also provide a written cost estimate for a training program.

4. *Training aircraft:* Obtaining a suitable training aircraft is perhaps the biggest variable to be considered. Two major questions are to be answered: How well is the aircraft in question maintained, and are you insured? If the aircraft is owned by a private flying club, your instructor should be a designated club flight instructor (on the club's insurance records) and you should be a club member in good standing. If the aircraft is to be rented from an individual, make certain you are insured under the owner's policy—take nothing for granted. Insurance companies charge extra for student training. If possible, have your name added to the aircraft owner's insurance

policy as a pilot of the aircraft to be used. This removes doubt as to being properly insured—assuming, of course, the owner has paid his premiums on time.

5. *Track record:* Perhaps the single most descriptive indicator of quality is the track record of the freelance instructor. How many students has he or she recommended for flight test? How many have passed the first time? A record of 80% passing their flight test on the first try is acceptable. Less than 80% is questionable as to the quality of the program being given by the instructor.

SELECTING YOUR TRAINING AIRCRAFT

High wing, low wing? T-tail or conventional? Cessna, Beech, Piper, or Grumman American? Indeed, there is a choice to be made. You are going to spend many hours in the cockpit, so why not choose the aircraft that you feel will best satisfy your desires. In practice, there are four basic choices: the Cessna 152, the Beech Skipper, the Piper Tomahawk, and the Grumman American Trainer.

The Cessna 152 is clearly acknowledged as the world's leading trainer; over 75% of all student pilots train in Cessna 152 type equipment (Figs. 1-1, 1-2). The aircraft is indeed a superb trainer—docile, forgiving, but demanding of good technique to truly perform a maneuver well. Without question, the Cessna 150/152 series has the finest and most effective system of flaps on any light plane. (The same is true for the 172/182 series of Cessna aircraft.) Being semi-Fowler flaps, they enhance lift as well as add significant drag when fully extended, thus becoming very effective air brakes.

Fig. 1-1. Cessna 152, the world's most popular trainer. Approximately 75% of all student pilots train in a Cessna 152.

Fig. 1-2. The instrument panel of a Cessna 152 is adequate for private and a large percentage of commercial pilot training.

The 152 is capable of all required flight maneuvers for the private or commercial pilot training program, including spins. If you weigh much over 200 pounds, however, it may be necessary that you train in a Cessna 172. The cockpit of the 152 is small and a bit crowded for two large people. The 172 handles much like a 152 but is much roomier inside, being a four-place aircraft.

If your leanings are toward pursuit-type aircraft, then a low-wing trainer may be more to your liking. The Beech Skipper may well fill the bill; it's a rugged machine built with typical Beech quality and thoroughness (Figs. 1-3, 1-4). Still another may be the Piper Tomahawk, a similar low-wing T-tail design. But how do they compare? Are they faster than the Cessna 152? Why the T-tail?

Many words have been written on the subject of the "best" aircraft configuration. The question to be answered is *the best for what purpose?* Our concern is with training aircraft so let us proceed with that in mind.

High-wing/low-wing: In reality, it's a matter of your personal preference. All four of the trainers noted above have about the same maximum speed and ceiling ratings. This is simply because all have similar engine horsepower ratings and approach/stall speeds. The latter is an FAA requirement which tends to standardize performance between classes of aircraft. If you like pursuit-type aircraft, choose a low-wing trainer, but don't expect to outrun your high-wing compatriot.

T-tail: The T-tail is a definite advantage to the high-speed turboprop and jet aircraft but of negligible benefit to trainers. Because the T-tail stabilizer operates in "undisturbed" air, it can be made smaller in frontal area than a conventional tail, thereby reducing parasite drag and permitting an increase in speed. At the low speeds of the trainer, the theoretical advantage of this is so small as to be of no real value. In reality, the T-tail has one major *disadvantage*. In making a soft field takeoff, the prop blast over a conventional tail is used to raise the nosewheel almost immediately at the start of the takeoff roll. Not so with the T-tail. The T-tail aircraft must gather adequate forward velocity for the stabilizer to become effective before the nosewheel can be raised. For training aircraft, the T-tail is more a matter of appearance than a functional advantage (Fig. 1-5).

Landing float: All low-wing aircraft tend to float a bit more than high-wing aircraft during flare and landing due to ground effect—not an undesirable characteristic, simply a fact with low-wing aircraft.

Flap effectiveness: First prize for effective flaps goes to the Cessna 152; last place to the Grumman American Trainer.

Spins: The Grumman American Trainer is prohibited from intentional spins, a poor characteristic for any training aircraft. Granted, spins are not a required part of the training curriculum for

Fig. 1-3. If your interest is in low-wing aircraft, the Beechcraft Skipper is an example of a clean design.

Fig. 1-4. To be of value as a trainer, the cockpit must feature a full panel as shown for the Beechcraft Skipper. Learning is accomplished quicker and more thoroughly using a full panel from the beginning. After full panel training has been mastered, only then can the student be expected to fly partial panel, for which a higher proficiency and workload is required.

either the private or commercial certificate, but what about the beginning student who simply makes a mistake at the wrong time during stall practice?

Taxi steering: Most tricycle aircraft employ nosewheel steering from the rudder pedals for directional control during taxi. Not so with the Grumman American Trainer, however. This aircraft employs a free-castering nosewheel and depends upon differential braking by the pilot for steering during taxi. Not necessarily a bad feature but certainly non-standard.

Fuel management: Cessna's high-wing gravity feed system is without a doubt the simplest. All low-wing aircraft require mechanical pumps to provide the engine with fuel.

All things considered, any of the four popular trainers listed herein are capable of servicing your basic flight instruction needs. The choice is really the one that appeals to you, given consideration of the tradeoffs noted above. From my standpoint, the preference is the Cessna 152 or Beech Skipper first, followed by the Piper Tomahawk and lastly the Grumman American Trainer.

GETTING YOUR STUDENT PERMIT

To be certificated as a pilot, you must meet the minimum physical standards and aeronautical skill requirements listed in the

11

Federal Aviation Regulations. Three steps are involved in the process. First, you must take a physical examination given by an FAA aviation medical examiner. The purpose of this examination is to assure that no defects exist which may be detrimental to flight. Second, you must take a written examination as a measure of your theoretical aeronautical knowledge. Third, you must demonstrate your flight proficiency by a flight test. In this chapter, we shall consider only the medical requirement. Chapter 4 is devoted to details of the written examination and flight training/flight testing.

As the very first step in learning to fly, obtain your student permit by taking your flight physical. The student permit is a part of the physical examination form which you will fill out when you take your exam. You must have a student permit and at least a Class III physical examination to solo. But why wait; get the examination out of the way before starting your flight program, thus assuring both yourself and your flight instructor that all is okay to proceed. The cost of the flight physical varies, depending upon the rates of the individual medical examiner. Typically, the charge for a Class II or III physical is from $30 to $50, depending somewhat on your age and general physical condition.

Medical Examination

An FAA physical may be given only by an aviation medical

Fig. 1-5. The Piper Tomahawk sports a T-tail—modern to look at but of little practical value for a trainer aircraft.

examiner who is specifically designated for the purpose. Your family doctor may or may not be so designated. To obtain a list of qualified medical examiners in any area, contact the FAA regional director or the FAA General Aviation District Office. Most fixed base operators, flight schools, Flight Service Stations, or state aviation authorities can provide a list of designated aviation medical examiners.

A medical certificate may be issued to an applicant who does not meet the medical standards required by regulation in certain instances. For example, an applicant who is moderately color blind may be able to read the standard aviation red, green, and white signals issued by a tower even though the medical examiner's cards were indistinct as to their color. In a situation of this nature, the applicant may apply to the FAA for a practical test of color blindness. A typical test is simply that of the applicant and a local FAA representative "reading" a group of light signals from a tower. If the applicant can successfully determine the color of the signals, his operating limitation may be removed. However, should he be unable to do so, it is likely that he will be prohibited from flight at night.

Any person who is denied a medical certificate may, within 30 days after the date of denial, apply in writting (in duplicate) to the Federal Air Surgeon, (attention Chief Aeromedical Certification Branch, Civil Aeromedical Institute, Federal Aviation Administration. P. O. Box 25082, Oklahoma City, Oklahoma 73125) for reconsideration of that denial. If such action is contemplated by an applicant, it is important that it be accomplished within 30 days after the date of the applicant's medical certificate denial. Following review by the FAA medical authorities, the denial may be upheld, the applicant may be asked to supply additional medical evidence, or simply to demonstrate competence of airmanship (as in the case of a physical handicap). Assuming the latter, the FAA may give a combined medical-private or medical-commercial flight test. If the pilot demonstrates competence, his certificate may be renewed without waiver; that is, additional "medical" flight tests will not be required. Many pilots are flying who are restricted to wheelchairs in their everyday life. Similarly, there are pilots legally flying who have vision in only one eye.

Medical Standards

Medical standards are divided into three classes. To qualify for a student or private pilot certificate, an applicant must pass the requirements for a third class medical certificate. A third class medical certificate is valid for a period of 24 calendar months and represents the minimum physical standards set forth by regulation. The certificate expires on the last day of the calendar month in which it was issued. For example, if a private pilot obtained a third class medical certificate on November 15, 1982, the certificate would expire at midnight on November 30, 1984.

A second class medical certificate is required for commercial pilots and flight instructors. This certificate denotes slightly higher minimum physical requirements—as should logically be expected of pilots involved in flight for hire. A second class medical certificate is valid for a period of 12 calendar months after the date of issue, whereupon it serves as a third class medical certificate for an additional 12 calendar months. Therefore, an applicant who qualified for a second class medical certificate on November 15, 1982, may render commercial aviation services until November 30, 1983. Should he wish to continue commercial operations, it will be necessary that he renew his second class medical prior to this date. However, should he wish to fly only as a private pilot, he may do so until November 30, 1984, at which time he must renew his medical if he is to continue flying.

The first class certificate is the most rigorous of the FAA physical examinations. This certificate is intended primarily for airline transport pilots. A first class certificate is valid for a period of six months as a first class certificate, six more months as a second class certificate, and twelve additional months as a third class certificate (24 months total).

The medical certificate is a required companion to the pilot certificate. Both must be carried at all times when performing the duties of a pilot.

Third Class Medical Certificate

To be eligible for a third-class medical certificate, an applicant must have distant visual acuity of 20/50 or better in each eye separately, without correction; if the vision in either or both eyes is poorer than 20/50 and is corrected to 20/30 or better in each eye with corrective glasses, the applicant may be qualified on condition he wear glasses while acting as an airman. He must be able to hear the whispered voice at three feet, have no acute or chronic diseases of the internal ear or disturbances in equilibrium. In addition, there must be no established medical history or diagnosis of personality disorders, psychosis, alcoholism, drug dependency, epilepsy, unexplained loss of consciousness, convulsive disorders, or other such items. The applicant must have no established medical history of myocardial infarction, angina pectoris, or other evidence of coronary heart disease that may reasonably be expected to lead to serious heart problems. A history of diabetes that requires insulin or similar agents for control may be disqualifying.

Fundamentally, the third class medical is a conventional physical examination whose purpose is to determine possible organic, functional, or other such defects which may make an applicant unable to safely perform the duties of an airman. A further purpose of the certificate is to give reasonable assurance that such will continue to be the case for a period of two years.

Second Class Medical Certificate

The second class medical, a requirement for a commercial pilot certificate, is similar to a third class certificate except for tolerances being somewhat tighter. To be eligible for a second class medical certificate, an applicant must have distant visual acuity of 20/20 or better in each eye separately without correction, or at least 20/100 in each eye separately corrected to 20/20 or better with corrective glasses, in which case the applicant may be qualified provided he wears glasses while exercising the privilege of his airman certificate. In addition to normal fields of vision and the ability to distinguish aviation red, green, and white, the applicant must have certain other eye requirements related to bifoveal fixation and vergence-phoria. An applicant for a second class medical certificate must be able to hear the whispered voice at eight feet with each ear separately.

Blood pressure limits are specified only for the first class medical certificate. However, as a general guide for the second and third class medical certificates, a maximum reading of 170/100 is a typical limit. The applicant's age, weight, and total physical condition is considered in cases of elevated blood pressure readings. Other requirements are similar to those of the third class certificate.

First Class Medical Certificate

The first class medical, needed for the Airline Transport Pilot certificate, embodies all of the requirements of the second class certificate with additional emphasis on sight, hearing, and heart condition. An applicant for a first class medical must have distant

Table 1-1. First Class Medical Blood Pressure Limits.

Age Group	Maximum readings (reclining blood pressure in mm)		Adjusted maximum readings (reclining blood pressure in mm)	
	Systolic	Diastolic	Systolic	Diastolic
20-29	140	88	---	---
30-39	145	92	155	98
40-49	155	96	165	100
50 and over	160	98	170	100

visual acuity of 20/20 or better in each eye separately without correction, or at least 20/100 in each eye separately corrected to 20/20 or better with corrective glasses, near vision of at least V = 1.00 at 18 inches with each eye separately, normal color vision, normal field vision, and no acute or chronic conditions of either eye that might interfer with the applicant's ability to perform his function as a pilot. The wearing of glasses is permitted. The applicant must have the ability to hear the whispered voice at a distance of at least 20 feet with each ear separately, or demonstrate a hearing acuity of at least 50% of normal in each ear throughout the effective speech and radio range as shown by a standard audiometer. After age 35 an applicant may be subject to an electrocardiographic examination to determine his heart condition. Blood pressure limits are shown in Table 1-1.

YOUR FLIGHT INSTRUCTOR

With this completion of your physical and the selection of your flight training program and your favorite training aircraft, you have gotten started in this flying business. Now the matter of progress rests between you and your flight instructor.

As a person, your flight instructor has between $10,000 and $20,000 invested in his CFI certificate. His or her training has been costly indeed. Legally, he is entirely responsible for your safety every time he acts in the capacity of a flight instructor. Furthermore, the FAA keeps a record of the number of students that successfully pass or fail their flight checks and the name of the instructor involved. Should there be evidence of too many failures associated with a particular instructor, the instructor is subject to re-examination by the FAA in order to keep his CFI certificate. In short, he or she has a significant investment to protect, in addition to undertaking the legal responsibilities associated with the training of a student. Do not expect your instructor's sign-off to solo, or take a cross-country, or a flight examination until such time as you are *truly* prepared. Your satisfactory performance is a measure of his ability to teach, and that is a necessary item to keeping one's CFI certificate. Simply remember that your instructor is as interested in your success as a pilot as you are. After all, your instructor is, in the long run, measured by *your* achievements.

It is necessary that a student and an instructor have a good comprehension of the abilities of each other and communicate well accordingly, particularly in the cockpit—the world's worst classroom. Upon occasion, a student and an instructor simply cannot communicate well. The reason generally is lack of a common experience base. Doctors and engineers, as student pilots, communicate best with instructors who can express flight facts in scientific terms, a language familiar to the doctor or engineer. Conversely, the businessman or the housewife generally prefer flight facts be explained in everyday language. The police officer or military student expect an element of discipline in their training program. Simply stated, flight students want new knowledge but they want such in a language with which they are already familiar. It is for this reason that occasionally a communications gulf develops between a student and an instructor. When this happens it is best to pair with another instructor in the hope of finding a more acceptable level of common experience. Established thus, training will continue efficiently. Your instructor is very aware of the benefits of a common experience level during training and he may himself suggest another instructor who would better suit your specific learning needs.

Chapter 2
Aviation Safety

The most dangerous part of flying is the drive to and from the airport. Aviation is still newsworthy but characteristic of the news media, only bad news or the bizarre rate headlines. In comparison to other forms of transportation, aviation accidents are publicized entirely out of proportion. Therefore, an unfortunate view of aviation safety is presented to the general public by the media. No wonder the beginning student often feels apprehensive about his or her safety.

"Is general aviation safe? Will my learning to fly jeopardize my family? What must I do to be a safe pilot? Should I take out more life insurance?" These and other like questions are the result of news media preconditioning. Perhaps they are not of significant concern to the pilot candidate, but are often the views of a spouse. Whatever the case may be, an answer must be forthcoming for flight training to proceed in an optimum fashion. Aviation safety must be understood; doubts must be erased.

At this point it would be simple to flood you with an avalanche of statistical data as collected, analyzed, and compared since the very beginning of flight. Statistical safety analyses have been generated for every facet of flight. Every accident, every cause factor, every related factor are matters of record. Study results and trends are all available in exacting statistical form (Fig. 2-1, Table 2-1). But will a listing of numbers, percentages, and conditions satisfy your anxiety, your personal questions about flight safety? Probably not. Simply adding statistics to your knowledge that reflect the behavior

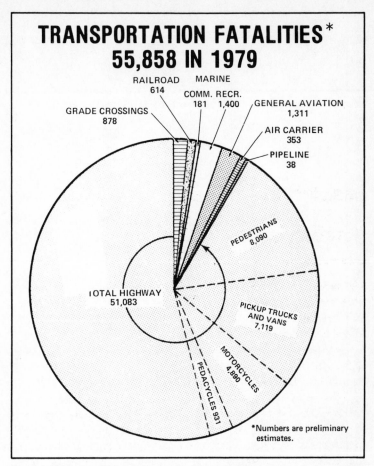

Fig. 2-1. One way of comparing types of transportation accidents. The safety statistics clearly show that flying is safer than walking!

of the flying community as a whole does not answer the question "What about *me?*"

On this basis, let's leave the statistics of aviation safety to the latter part of this chapter and proceed directly to the question at hand. *What about you?*

As a start, recognize that some 90% of all aviation accidents are attributable to the pilot. Only 10% can be blamed on the flying machine. The first major point in the answer to the question is simply that the primary cause of aviation accidents is you—the pilot. This is an encouraging fact because you, as a pilot, can do a

Table 2-1. A Ten-Year Record of U.S. General Aviation Accidents. Note How the Average Accident Rate has Decreased Over the Past Ten Years.

Accidents, Fatalities, Rates
U.S. General Aviation
1970-1979

| | Accidents | | Fatalities | Aircraft-Hours Flown (000) c/ | Aircraft-Miles Flown (000) c/ | Accident Rates | | | |
| | Total | Fatal | | | | Per 100,000 Aircraft-Hours Flown | | Per Million Aircraft Miles Flown | |
						Total	Fatal	Total	Fatal
1970	4,712[a]	641[a]	1,310	26,030	3,207,127[d]	18.1	2.46	1.47	0.200
1971	4,648	661	1,355	25,512	3,143,181	18.2	2.59	1.48	0.211
1972	4,256[a]	695[a]	1,426[b]	26,974	3,317,100	15.8	2.57	1.28	0.209
1973	4,255[a]	723[a]	1,412	29,974[f]	3,686,802[e]	14.2	2.41	1.15	0.196
1974	4,425[a]	729[a]	1,438	31,413[f]	3,863,799[e]	14.1	2.31	1.14	0.188
1975	4,237[a]	675[a]	1,345	32,024[f]	3,938,952[e]	13.2	2.10	1.08	0.171
1976	4,193[a]	695[a]	1,320	33,922[f]	4,172,406[e]	12.3	2.04	1.00	0.166
1977	4,286[a]	702[a]	1,436	35,792[f]	4,402,126[e]	12.0	1.96	0.97	0.159
1978	4,494[a]	793[b]	1,770[b]	39,400	4,964,400[e]	11.4	2.01	0.90	0.159
1979P	4,238	658	1,311	39,900	5,052,800[e]	10.6	1.65	0.84	0.130

[a] Suicide/sabotage accidents included in all computations except rates (1970-1, 1972-3, 1973-2, 1974-2, 1975-2, 1976-4, 1977-1, 1978-2).
[b] Includes air carrier fatalities (1972-5, 1978-142) when in collision with general aviation aircraft.
[c] Source: FAA.
[d] Beginning in 1970, the decrease in aircraft-miles flown is the result of a change in the standard for estimating miles flown.
[e] Estimated by NTSB.
[f] Revised by FAA.

P Preliminary

great deal to enhance your personal safety in flight. By and large, *you* are directly responsible for and in charge of *your* safety. It can be at whatever level of acceptability that *you* want it to be. Think not? Then read on and consider the following leading causes of aviation accidents. In each of these instances, the accident rate could drop to zero if you, the pilot, want it to. The *true* accident is rare—the one in which the pilot finds himself involved but had nothing to do with the circumstances. So very many accidents—the majority, in fact—are entirely preventable. Safety is, for all practical purposes, *what you want it to be.*

WEATHER-RELATED ACCIDENTS

Weather-related accidents are clearly the leading type of general aviation accident which results in fatalities. Approximately one-third of all aviation fatalities are due to weather in one way or another. These accidents occur with disturbing regularity despite improvements in aircraft, instrumentation, training, training facilities, the air traffic control system, weather facilities, weather services, and navigational aids.

The trend of fatal weather-involved general aviation accidents has been increasing steadily, while the trend of the accident rate per 100,000 aircraft-hours flown (all fatal accidents) has been generally downward.

Causal Factors

Pilots with fewer flight hours are more frequently involved in weather accidents, especially those pilots with more than 100 but less than 300 total flight hours. Perhaps the explanation for the peak is that by the time a pilot has accumulated 100 to 299 hours, he is confident of his flying ability, even though his actual flying experience is low. His or her experience with flying in a variety of adverse weather conditions is, of course, even lower. Therefore, less experienced pilots might not be aware of the potential dangers involved with adverse weather.

The most frequently cited cause of weather-related accidents is continued VFR into adverse weather conditions. The next most frequently cited causes (by order of frequency), are inadequate preflight preparation and/or planning, attempted operation beyond experience/capability level, and failure to obtain/maintain flying speed.

Almost 62 percent of the pilots involved in weather-related

accidents do not file flight plans. Based on these statistics, it is possible that there may be a relationship between accident involvement and lack of a filed flight plan.

Low ceiling is the most frequently cited weather phenomenon in weather-involved fatal general aviation accidents. Fog and rain are the next two most frequently cited phenomena. Although such phenomena as turbulence, thunderstorm activity, and icing are involved in a significant number of cases, the numbers are small compared to those of low ceilings and fog.

From a case study of weather-related accidents, it was determined that 74 percent of the National Weather Service forecasts were considered to have been either substantially correct, or the weather was *better* than predicted. On the other hand, 11 percent were considered to have been worse than forecast. Of those, in only about 5.5 percent of the cases, was the forecast completely inaccurate, or the weather considered to have been considerably worse than forecast.

If you can assume that ¾ of the forecasts you receive will be reasonably accurate, you cannot ignore the forecasts when planning a flight. Experienced pilots are aware that forecasts cannot be considered "gospel," but they also know that they cannot be ignored. Forecasts should be treated as the best professional advice available.

Weather and Your Flying Safety

Can your weather-related flying safety be controlled to a level acceptable to you? Yes indeed! How? Through adherance to a few common-sense flying practices. Experience has taught the following as excellent habits to adopt to extend one's longevity as a pilot;

1. Always obtain a weather briefing before flight. During flight, keep a radio tuned to FSS or Flight Watch for changes in the weather scene.

2. File a flight plan! An experience of the author's a number of years ago taught the value of a flight plan as a weather-warning tool. On an extended cross-country. I had departed Albuquerque, New Mexico, for a flight across the mountains to Phoenix, Arizona. Although it was about 4:00 p.m. when I filed my flight plan, Albuquerque FSS did not indicate any of the usual afternoon buildups en route. About 40 miles west of Albuquerque VOR, I heard FSS calling my aircraft number. I quickly replied and was advised that a pilot report just received noted heavy thunderstorm activity at St.

Johns (about the halfway point to Phoenix). Not wanting to encounter thunderstorms, turbulence, mountains, the onset of darkness, and a high density altitude all at the same time, I returned to Albuquerque for dinner and a relaxed evening. The flight to Phoenix the following morning was superb, crystal clear, smooth—an exhilarating experience. Had I not filed a flight plan, the pilot report of St. John's weather would not have reached me. If one is to be weathered in anyway, it might as well be at Albuquerque. There are few alternate choices en route to Phoenix.

3. When marginal weather is known to prevail, *always* plan a way out *before* beginning a flight. Encountering adverse weather itself keeps a pilot busy just flying the aircraft without having to search through maps for an alternate route, other VOR frequencies, and the like.

4. If you are a VFR-only private pilot, *absolutely do not attempt* instrument flight! The few hours of hood time contained in your training is purely for purposes of instrument *familiarization* to support your VFR abilities. Statistics clearly show that the VFR pilot who enters IFR conditions has, as a maximum, *eight minutes* to live! Know your weather limitation and those of your aircraft and abide by them accordingly.

5. Reject the pressures and advice of well-meaning but untrained passengers with go-home-itis. *You* are the pilot in command. *You* are responsible for the safety of all. Eternity is such a long time—why rush it?

6. Study weather, visit your FSS, and learn what weather is all about. Understand the sky and its moods. There are many excellent books on the subject. A few dollars and a few hours is all that it takes.

SAFETY AND YOUR HEALTH

Just as your aircraft is required to undergo regular checks and maintenance, you are also required to undergo regular medical examinations to ensure your fitness to fly. The physical standards you are required to meet are minimum standards. You do not have to be a superman to fly. Many defects can be compensated for as, for example, wearing glasses for visual defects. You may be required to demonstrate by a medical flight test that you can compensate for any other defects of potential significance to flight safety.

Student pilots should visit a Designated Aviation Medical Examiner and determine if they meet the standards *before* spending much money taking flying instructions.

It should be recalled that humans are essentially earthbound creatures. However, if we are aware of certain aeromedical factors and pay attention to them, we can leave the earth and fly safely. What follows will not be a comprehensive lesson in aviation medicine. It will simply point out the more important factors with which you should be familiar prior to flying.

Modern industry's record in providing reliable equipment is very good. When the pilot enters the aircraft, he becomes an integral part of the man machine system. He is just as essential to a successful flight as the control surfaces. To ignore the pilot in preflight planning is as senseless as failing to inspect the integrity of the control surfaces or any other vital part of the machine. The pilot himself has the sole responsibility for determining his reliability prior to entering the cockpit for flight.

While piloting an aircraft, an individual should be free of conditions which are harmful to alertness, ability to make correct decisions, and rapid reaction times. Persons with conditions which are apt to produce sudden incapacitation (such as epilepsy, serious heart trouble, uncontrolled diabetes mellitus or diabetes mellitus requiring hypoglycemic agents, and certain other conditions hazardous to flight) cannot be medically certified according to the Federal Aviation Regulations. Conditions such as acute infections, anemias, and peptic ulcers are temporarily disqualifying. Consult your Aviation Medical Examiner when in doubt about any aspect of your health status, just as you would consult a licensed aviation mechanic when in doubt about the engine status.

Fatigue

Fatigue generally slows reaction times and causes foolish errors due to inattention. In addition to the most common cause of fatigue (insufficient rest and loss of sleep), the pressures of business, financial worries, and family problems can be important contributing factors. If your fatigue is marked prior to a given flight, *don't fly*. To prevent fatigue effects during long flights, keep active with respect to making ground checks, radio-navigation position plotting, and remaining mentally active. engine noise is a definite cause of fatigue. The use of a headset is remarkably effective in reducing fatigue. Earplugs are also beneficial.

Hypoxia

Hypoxia is in simple terms a lack of sufficient oxygen to keep the brain and other body tissues functioning properly. Wide individual variation occurs with respect to susceptibility to hypoxia. In addition to progressively insufficient oxygen at higher altitudes, anything interfering with the blood's ability to carry oxygen can contribute to hypoxia (anemias, carbon monoxide, and certain drugs). Also, alcohol and various drugs decrease the brain's tolerance to hypoxia.

Your body has no built-in alarm system to let you know when you are not getting enough oxygen. It is impossible to predict when or where hypoxia will occur during a given flight, or how it will manifest itself.

A major early symptom of hypoxia is an increased sense of well-being (referred to as *euphoria*). This progresses to slowed reactions, impaired thinking ability, unusual fatigue, and a dull headache feeling.

The symptoms are slow but progressive, insidious in onset, and are most marked at altitudes above ten thousand feet. Night vision, however, can be impaired starting at altitudes lower than ten thousand feet. Heavy smokers may also experience early symptoms of hypoxia at altitudes lower than non-smokers.

If you observe the general rule of not flying above ten thousand feet without supplemental oxygen, you will not get into trouble.

Alcohol

Do not fly while under the influence of alcohol. An excellent rule is to allow 24 hours between the last drink and takeoff time. Even small amounts of alcohol in the system can adversely affect judgment and decision-making abilities. Remember that your body metabolizes alcohol at a fixed rate, and no amount of coffee or medication will alter this rate.

By all means, do not fly with a hangover or a "masked hangover" (symptoms suppressed by aspirin or other medication). Every year 40 to 50 fatalities occur as a direct result of flying under the influence of alcohol. This type of accident is needless!

Drugs

Self-medication or taking medicine in any form when you are flying can be extremely hazardous. Even simple home or over-the-counter remedies and drugs such as laxatives, tranquilizers, and appetite suppressors may seriously impair the judgment and coordination needed while flying. This is often due to a synergistic reaction between the drug and the reduced supply of oxygen while at altitude. The safest rule is to take *no* medicine while flying, except on the advice of your Aviation Medical Examiner. It should also be remembered that the condition for which the drug is required may in itself be hazardous to flying, even when the symptoms are suppressed by the drug.

Certain specific drugs which have been associated with aircraft accidents in the recent past are *antihistamines* (widely prescribed for hay fever and other allergies); *tranquilizers* (prescribed for nervous conditions, hypertension, and other conditions); *reducing drugs* (amphetimines and other appetite suppressing drugs can produce sensations of well-being which have an adverse effect on judgment) ; *barbituates, nerve tonics* or *pills* (prescribed for digestive and other disorders, barbituates produce a marked suppression on mental alertness).

Vertigo

The word vertigo itself is difficult to define. To earthbound individuals, it usually means dizziness or swimming of the head. To a pilot it means, in simple terms, that he doesn't know which end is up. In fact, vertigo during flight can have fatal consequences.

On the ground, we know which way is up by the combined use of three senses:

☐ *Vision:* We can *see* where we are in relation to fixed objects.

☐ *Pressure:* Gravitational pull on muscles and joints tells us which way is down.

☐ *Special parts in our inner ear:* The otoliths tell us which way is down by gravitational pull.

It should be noted that accelerations of the body are detected by the fluid in the semicircular canals of the inner ear, and this tells us when we change position. However, in the absence of a visual reference (such as flying into a cloud or overcast), the accelerations can be confusing, especially since their forces can be misinterpreted as gravitation pulls on the muscles and otoliths. The result is often disorientation and vertigo (or dizziness).

All pilots should have an instructor pilot produce maneuvers which will produce the sensation of vertigo, or arrange to ride the Vertigon trainer, if available. Once experienced, later unanticipated incidents of vertigo can be overcome. Closing the eyes for a second or two may help, as will watching the flight instruments, believing them, and controlling the airplane in accordance with the information presented on the instruments. All pilots should obtain the minimum training recommended by the FAA for attitude control of aircraft solely by reference to the gyroscopic instruments.

Pilots are susceptible to experiencing vertigo at night, and in any flight condition when outside visibility is reduced to the point that the horizon is obscured. An additional type of vertigo is known as *flicker vertigo*. Light, flickering at certain frequencies, from four to twenty times per second, can produce unpleasant and dangerous reactions in some persons. These reactions may include nausea, dizziness, unconsciousness, or even reactions similar to an epileptic fit. In a single-engine propeller airplane, heading into the sun, the propeller may cut the sun to give this flashing effect, particularly during landings when the engine is throttled back. These undesirable effects may be avoided by not staring directly through the prop for more than a moment, and by making frequent but small changes in rpm. The flickering light traversing helicopter blades has been known to cause this difficulty, as has the bounce-back from rotating beacons on aircraft which have penetrated clouds. If the beacon is bothersome, shut it off during these periods.

Carbon Monoxide

Carbon monoxide is a colorless, odorless, tasteless byproduct of an internal combustion engine and is always present in exhaust fumes. Even minute quantities of carbon monoxide breathed over a long period of time may lead to dire consequences.

For biochemical reasons, carbon monoxide has a greater ability to combine with the hemoglobin of the blood than oxygen. Furthermore, once carbon monoxide is absorbed in the blood, it sticks like glue to the hemoglobin and actually prevents the oxygen from attaching to the hemoglobin.

Most heaters in light aircraft use air flowing over the manifold. So if you have to use the heater, be wary if you smell exhaust fumes. The onset of symptoms is insidious, with "blurred thinking," a possible feeling of uneasiness, and subsequent dizziness. Later, headache occurs. Immediately shut off the heater, open the air ventilators, descend to lower altitudes, and land at the nearest airfield. Consult an Aviation Medical Examiner. It may take several days to fully recover and clear the body of the carbon monoxide.

Vision

On the ground, reduced or impaired vision can sometimes be dangerous, depending on where you are and what you are doing. In flying, it is always dangerous. On the ground or in the air, a number of factors such as hypoxia, carbon monoxide, alcohol, drugs, fatigue, or even bright sunlight can affect your vision. In the air, these effects are critical.

Some good specific rules are: Make use of sunglasses on bright days to avoid eye fatigue. During night flights, use red covers on the flashlights to avoid destroying any dark adaptation. Remember that drugs, alcohol, heavy smoking, and the other factors mentioned have early affects on visual acuity.

Scuba Diving

You may use your plane to fly to a sea resort or lake for a day's scuba diving, and then fly home, all within a few hours time. This can be dangerous, particularly if you have been diving to depths for any length of time.

Under the increased pressure of the water, excess nitrogen is absorbed into your system. If sufficient time has not lapsed prior to takeoff for your system to rid itself of this excess gas, you may experience the bends at altitudes under 10,000 feet where most lightplanes fly. A nitrogen renormalization period of 24 hours is recommended before the beginning of a flight.

Panic

The development of panic in inexperienced pilots is a process which can get into a vicious circle and lead to unwise and precipitous actions. If lost, or in some other predicament, forcibly take stock of yourself and do not allow panic to mushroom. *Panic can be controlled*. Remember, *prevent panic to think straight*. Fear is a normal protective reaction and occurs in normal individuals. Fear progression to panic is an abnormal development.

For Additional Information

If you'd like additional or more specific information concerning the preceding or other aeromedical flight factors, direct inquiries to the Federal Air Surgeon, Federal Aviation Agency, Washington, D.C. 20553, or, if geographically more convenient, to the Director, Civil Aeromedical Institute, P. O. Box 25082, Oklahoma City, Oklahoma, 73125. Also, see *Pilot's Aeromedical Guide,* TAB Book No. 2287.

TRAINING AND YOUR SAFETY

Regretably, some candidate pilots treat flight training as a high school course—just make a 70, that's all. No use straining one's self. After all, a 70 passes the course just as well as 100 does. The flight test? "Well, if I don't pass it the first time, I'll simply practice what I missed and try again."

Yes, it is possible to adopt a "70 attitude" and get a pilot certificate simply because the law must establish a minimum and that minimum is a 70. While the flight test does not provide a grade as such, it does make allowances for less-than-ideal performance; it asks only for an average level of proficiency to meet the prescribed standards of performance. But what about your flight experiences following the award of your certificate? Will they likewise tolerate a 70? The accident records clearly say *no!* Let's consider the following lack of training or lack of common-sense types of accidents.

Stall/Spin

Running a close second to weather-related accidents, the stall/spin is an altogether too common type of accident. The scenario usually proceeds as follows: Approaching an unfamiliar airport, the pilot throttles back to slow the aircraft. Next he becomes entirely preoccupied with the airport—where is the windsock, the taxiway, aircraft parking, etc.? His speed on downwind continues to decrease, unnoticed. Now a turn to base; the wing blocks the view of the airport, so raise the wing and add opposite rudder. Then it happens—so very quickly, so entirely decisive—the stall, the spin, and an insufficient altitude in which to recover. Another accident, another headline. But why?

The analysis of this type of accident is as classic as the accident itself. The pilot simply failed to watch his inside references—his instruments—while slowing for an approach. Airspeed, altitude, the turn coordinator ball, and even the stall warning horn were all neglected. The victim became so engrossed in his study of the outside scene, the unfamiliar airport and its surroundings, that management of the aircraft was left entirely to feel and guesswork. Under these circumstances, a pilot will *invariably* pull back on the aircraft control column. As power has already been reduced in preparation for a landing, pulling back on the control column simply reduces speed even more. The aircraft next slides effortlessly into the region of reverse command awaiting that final application of back pressure which results in a stall. If the rudder is being misused (that is, the aircraft is being flown in an uncoordinated manner), the spin necessarily follows.

A skid is by far the most dangerous of the types of uncoordinated flight maneuvers. When a stall is incurred during a skid, the wing to the inside of the turn stalls first. The aircraft then abruptly snap rolls over the top and the nose drops straight down, waiting for the pilot to begin the recovery procedure. At the altitude of a traffic pattern, there simply isn't room left to effect a recovery in most aircraft, even if flown by a skilled aerobatic pilot.

What went wrong? Why did our victim make such a basic flying error? Was it lack of training? Lack of practice? Anxiety due to an unfamiliar set of surroundings? Perhaps all of the preceding played a part. In principle, however, our pilot victim broke a major commandment of flight, namely, "Thou shalt fly an aircraft by using *both* inside and outside references." One set alone is not enough.

When instructing a student in ground reference maneuvers, pylon eights, and traffic patterns, a major objective of the instruction is to insist the maneuvers be done accurately. Only by using *both* inside and outside references can this objective be met. The purpose of this is therefore to install in the student a discipline, an unforgettable habit to always use *both* sets of references in flying. The integration of instrument readings with visual perceptions is basic to *all* phases of flight—takeoff, cruise, night flight, landing, turning, etc. The flight instructor offers the training but only the student can develop the desired habit pattern. The dedicated student does so invariably. But what about that 70% passing grade-type student? If he flys long enough, that other 30% of the possible flight situations will eventually present themselves for solution. What then? The stall/spin accident is indeed a testimony so that unlearned 30%—a totally unnecessary kind of accident that simply denotes poor acceptance of training on the student pilot's part. This kind of accident needs never happen—its prevention is entirely up to you.

Fuel Mismanagement

PLANE LANDS SHORT OF RUNWAY, FOUR KILLED, FUEL TANKS DRY—so reads the headlines. The circumstances are repeated year after year. Fuel exhaustion continues to be a major cause of accidents. In one instance, a pilot passed over *eleven* airports where fuel could be obtained only to fall short of his destination (the twelfth airport) for lack of fuel. Why this happens is a mystery to me, but it *does* happen year after year—another preventable type of accident.

Lightplane crashes, auxiliary tanks full, main tanks empty: Fuel mismanagement, concludes the accident report. Lack of knowledge on the pilot's part as to operation of the aircraft fuel-feed system is indeed an accident causative factor. This event does have some subtleties that compound the problem. For example, some aircraft take up to 20 seconds to start feeding fuel from an auxiliary tank after switching from the main tank if the pilot has run the main dry (a poor practice.) This leaves a long period of silence before power is again available. Another example deals with a popular make of aircraft that normally employs both tanks on takeoff. However, a few models of the design had a slightly different fuel system and require takeoff on one tank only. The moral to the story is *know thy aircraft fuel system.*

Another accident causative factor relating to fuel management is water found in the fuel. Poor preflight? Failure to keep tanks full? Poor fuel storage facilities? Yet another example of fuel mismanagement is the aircraft fueled with jet fuel by mistake. Poor preflight? Failure to observe the fueling operation?

Whatever the error, it is the pilot who must pay the price for he is responsible for safety of flight. If only more pilots would take the time to observe that their aircraft is correctly fueled, this type of accident would be significantly decreased. Simply remember, the next time you order your aircraft fueled by phone, that both the secretary taking the order and the line boy fueling your aircraft are minimum-wage employees. Frankly speaking, their jobs border on the routine and offer little promotion incentive. A waitress in a restaurant is awarded a 15% tip for good service, but not so your line boy, even though his service is vital to your health. The same applies to the secretary in taking of fuel orders and dispatching of the fuel truck. Seldom do either of these FBO service employees enjoy a simple "thank you" for their services. Yes, mistakes by line service personnel do occur. Prevent them by watching your aircraft being fueled—just as corporate and air carrier pilots do.

Other Preventable Accidents

The list of perfectly preventable accidents goes on and on. Just to name a few true incidents, we have:

☐ Low flying; pilot collides with electric wires.

☐ Pilot caught in canyon, unable to climb out due to high density altitude.

☐ Pilot fails to preflight aircraft; attempts takeoff without removing control locks.

☐ Pilot takes off dragging tiedown chain and stake from tail of aircraft; reports large trim change needed in flight.

☐ Pilot fails to notice rudder of aircraft is removed for repair; attempted takeoff results in immediate crash, passenger injured.

☐ Passenger opens Thermos of very hot coffee at high altitude; scalding spray strikes pilot in eyes causing temporary blindness. (The boiling point of a fluid is reduced with altitude. In this instance, the coffee literally exploded when the cap was removed.)

Unbelievable? Indeed, some accidents should be featured in *Believe It Or Not* or perhaps the *Guiness Book of World Records*. A little common sense is all that is required to prevent their happening.

In summary, a large percentage of aircraft accidents are truly preventable. The question is simply, "Will you allow yourself to become an accident statistic?" Safety of flight is really *up to you* in the long run. Aviation is not inherently dangerous but it is most unforgiving to the careless or the reckless.

THE NATIONAL TRANSPORTATION SAFETY BOARD

Aviation safety is a major concern to two federal agencies, namely the Federal Aviation Administration and the National Transporation Safety Board (NTSB). These two agencies operate in a cooperative fashion with regards to aviation safety (Fig. 2-2). Most pilots, including student pilots, are aware of the activities and responsibilities of the FAA. The NTSB, however, is often unknown to the average pilot, even though the fundamental mission of this agency is transportation safety. The next few paragraphs are intended to introduce you to the NTSB. Their actions are directly responsible for the steady improvement in aviation's safety record:

"The National Transportation Safety Board is an independent Federal agency that serves as the overseer of U.S. transportation safety; its responsibilities are intermodal in scope and include railroad, highway, pipeline, marine and civil aviation transportation.

"The mission of the Safety Board is to improve transportation safety. This is done primarily by determining the probable cause of accidents through direct investigations and public hearings, and secondarily through staff review and analysis of accident informa-

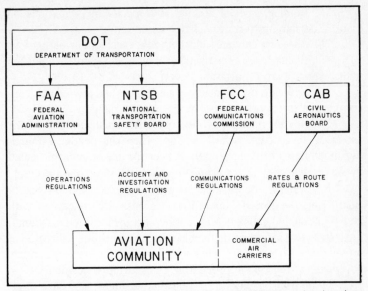

Fig. 2-2. The National Transportation Safety Board (NTSB) is on a level equivalent to that of the FAA. Their purpose is to investigate transportation accidents and determine the probable cause.

tion, through evaluations of operations, effectiveness, and performance of other agencies, through special studies and safety investigations, and through published recommendations and reports to Congress.

"Under the Independent Safety Board Act, the Board has the following authority:

"1. Investigate, determine the facts, conditions, and circumstances and determine the cause of probable cause of:

    a. Civil aircraft accidents;

    b. Railroad accidents in which there is a fatality, substantial property damage, or which involves a passenger train;

    c. Pipeline accidents in which there is a fatality or substantial property damage;

    d. Major marine casualties; and

    e. Highway accidents, including any railroad grade crossing accidents, that it selects in cooperation with the States.

"2. Issue safety recommendations, recommending corrective actions to reduce the likelihood of recurrence of transportation accidents;

"3. Conduct special studies and special investigations on matters pertaining to transportation safety;

"4. Assess techniques and methods of accident investigation;

"5. Evaluate the effectiveness of other government agencies with respect to transportation safety;

"6. Evaluate the adequacy of safeguards and procedures concerning the transportation of hazardous materials;

"7. Review on appeal orders of the Administrator of the FAA and the Commandant of the U.S. Coast Guard suspending, amending, modifying, or revoking operating certificates or licenses issued by these agencies." [*This includes all pilot certificates and ratings.*]

"In carrying out its mission under the Independent Safety Board Act, the Board's operations are basically concerned with investigation and cause determination of transportation accidents, the issuance of safety recommendations, special studies directed at preventing such accidents, and evaluating the effectiveness of other government agencies in preventing transportation accidents.

"In air transportation, the Board determines the cause of all civil aircraft accidents that occur in the United States, a function that cannot be delegated to any other department or agency. The Board investigates all air carrier accidents, most fatal lightplane accidents, and other selected accidents. The Board has, however, temporarily authorized the Federal Aviation Administration to investigate most non-fatal lightplane and helicopter accidents, but retains the statutory duty to determine probable cause in each case.

"The Board may hold a public hearing in which the facts determined in its investigation are presented as evidence and persons involved in the accident testify and are questioned. In such cases, the Safety Board establishes a Board of Inquiry, which is presided over by a Safety Board Member and witnesses are called to testify and be questioned under oath.

"Parties associated with the investigation are generally invited to participate in the public hearing process. In an aviation case, for instance, such parties would include the airline, the manufacturer of the aircraft and engines, the Federal Aviation Administration, and the pilot and air traffic controller unions. After the initial interrogation of witnesses by the Board, each party is also given the opportunity to question witnesses and, after the hearing is over, may submit their conclusions and recommendations to the Board.

"Such hearings are always public and are designed to help the Board determine the cause of an accident and provide information on which the Board may be able to develop recommendations to prevent other accidents. After all the evidence developed at the accident site and at the hearing is analyzed, a report setting forth the facts, conditions, and circumstances of the accident and the probable cause of the accident, is adopted by the Board and released to the public.

"All actions and decisions of the Safety Board, in the form of accident reports, special studies, statistical reviews, safety recommendations, and press releases are made public.

"Single copies of such documents are usually available free, although subscription rates apply on certain recurring publications. Various specific mailing lists are maintained by the Board to support this public distribution. Details on publications available may be obtained on request by writing to the Publications Section, National Transportation Safety Board, Washington, D.C. 20594.

"Safety recommendations are based on information developed from Board investigative findings and special studies and are aimed at preventing accidents and correcting unsafe conditions in transportation. As an official action of the Board, each recommendation is voted on by Board Members and signed by the Chairman. Recommendations are issued to Federal agencies, State governments, trade organizations, or companies, as appropriate. The Board's recommendations are not mandatory. However, as required by the Act, all are made public and they have a high rate of acceptance. Recommendations may be issued at any time, sometimes within a few days after an accident, to initiate quick action to prevent additional accidents.

"When the U.S. Department of Transportation receives a Board recommendation—and DOT receives more than 80 percent of all recommendations—the Secretary is required to reply within 90 days and set forth how they intend to respond to the recommendation. If the recommendation is rejected, in whole or in part, they are requested to explain the reasons. All responses are evaluated by the Board's technical staff. The five-member Board then makes the final determination as to whether the recommendation has been complied with adequately. As with the original recommendation, all responses and correspondence are made public.

"The Safety Board also undertakes special studies of safety problems in air and surface transportation. Much of the information used in such studies, whether technical or statistical in form, is acquired as the result of accident investigative findings. Subsequent analysis of such material provides the Board with the facts necessary to issue a special study delineating in detail problems or weaknesses in specific areas of transportation safety. Such studies frequently provide the basis for safety recommendations.

"During the first eleven years of its existence, the Safety Board issued 2800 recommendations—about 87 percent of which were accepted by Government and the transportation industry. This "acceptance rate" itself is an impressive testament since the Board's recommendations do not carry the force of law. In addition, they may reflect on the policies and procedures of other government agencies or private industry. In its mandate to the Board, Congress clearly emphasized the importance of the safety recommendation, and in the language of the Independent Safety Board Act of 1974 ordered the Board "to propose corrective steps to make the transportation of persons as safe and free from risk as is possible."

"Under the Board's policy, its safety recommendations must meet four criteria: clarity, conciseness, technical feasibility, and adequate support. These qualities ensure the effectiveness of the safety recommendations and facilitate its followup. Each recommendation designates the person or party expected to take action, describes the action the Board expects, and clearly states the safety need that is to be satisfied.

"The Board's recommendations have resulted in the Federal Aviation Administration's issuance of operations and engineering alert bulletins and notices, revising maintenance manuals, improving designs and instrumentation, airworthiness directives, and improvements in pilot-training programs. In addition, the recommendations produced improved aircraft crashworthiness and survival equipment requirements, and improvements in the air traffic control system and radar identification.

"Weather is a major factor in civil aviation accidents during all segments of flight—takeoff, en route, and approach and landing. As a result of the Safety Board's investigation into several weather-involved air carrier accidents, a total of 20 safety recommendations were issued. These recommendations concerned:

☐ Collection of thunderstorm data and the classification of the intensity of storms.

☐ Development and implementation of electronically-displayed weather on controller radarscopes.

☐ Detection and forecasting of wind shear.

☐ Training for controllers on the effect winds produced by thunderstorms can have on an airplane's flightpath control.

☐ Training for pilots in wind shear penetration.

☐ Dissemination and timeliness of weather information in the terminal area.

"Included also were recommendations for the development of airborne equipment which would alert the pilot to the need for rapid corrective measures and for airborne equipment which would permit a pilot to transition from instrument to visual reference without degradation of vertical guidance during the final segment of an instrument approach.

"These Safety Board recommendations have resulted in the development and deployment of wind shear detection systems, new general aviation airport wind indicators, additional education of pilots regarding operations in unfavorable wind conditions, and a program to improve and display weather and storm intensity on air traffic control radar scopes. In addition, the FAA has established common language to describe thunderstorm intensity and has improved the dissemination of severe weather information by revising the priorities of FAA Flight Service Stations and expediting the transmission of severe weather data in terminal areas. The FAA also plans to provide automated continuous severe weather broadcasts in the future."

Flying safety is indeed a major concern of the FAA, the NTSB, and your flight instructor. Aviation symposiums are devoted to the subject; flying magazines never fail to print article after article on the subject; books on flight safety abound. There is no lack of explanatory material. The elements of safety are widely known and readily distributed. In the final analysis, however, *safety is up to you*. Within the inherent limits of life itself, you may have as much safety as you wish.

Chapter 3
Learning

Without a doubt, one of the truly significant ways to save real dollars in obtaining your private or commercial certificate is to understand the learning process. The average person requires some 53 hours of flight training to obtain a private pilot's certificate. The legal minimum is 40 hours. Can the task be done in 40 hours? The answer is yes, *if* you really apply the principles of learning throughout your training. A savings of 13 hours can amount to a reduction in cost of up to $500—a sizable amount for simply increasing one's own learning efficiency.

Let's begin by developing an understanding of the learning process itself, then look at a few proven ways for applying learning basics as money-savers. The next few paragraphs, which describe the fundamentals of learning, have been paraphrased from the *Aviation Instructor's Handbook*, AC 60-14, the handbook your flight instructor is required to master before getting his instructor's certificate. Understand what your instructor is required to know about learning processes and you will better understand what he or she is trying to teach you. To illustrate how this fits into your flight training program, I have inserted explanatory comments. In short, let us understand both sides of the learning coin—the instructor's and yours.

THE LEARNING PROCESS

The ability to learn is one of humanity's most outstanding characteristics. Learning occurs continuously throughout a person's lifetime. To define learning, it is necessary to analyze what happens to the individual. As a result of a learning experience, and individual's way of perceiving, thinking, feeling, and doing may change. Thus, learning can be defined as *a change in behavior as a result of experience*. The behavior can be physical and overt, or it can be intellectual or attitudinal. Whatever the case, unless behavior is changed by a learning process, learning has not taken place. Your instructor seeks a behavioral change, one that can be measured by a written or a flight test to assure that learning has taken place.

Learning is an individual process. The instructor cannot do it for the student; knowledge cannot be poured into the student's head. The student can learn only from individual experiences.

"Learning" and "knowledge" cannot exist apart from a person. A person's knowledge is a result of experience, and no two people have had identical experiences. Even when observing the same event, two people react differently; they learn different things from it, according to the manner in which the situation affects their individual needs. Previous experience conditions a person to respond to some things and to ignore others.

All learning is by experience, but it takes place in different forms and in varying degrees of richness and depth. It is for this reason that your instructor prepares individual lesson plans designed to fit your learning characteristics.

THE LAWS OF LEARNING

The laws of learning are universal. Let's look at them from the instructor's viewpoint and then see how they fit you.

Law of Readiness: Individuals learn best when they are *ready* to learn, and they do not learn much if they see no reason for learning. Getting students ready to learn is usually the instructor's responsibility. If students have a strong purpose, a clear objective, and a well-fixed reason for learning something, they make more progress than if they lack motivation. Readiness implies a degree of singlemindedness and eagerness. When students are ready to learn, they meet the instructor at least halfway, and this simplifies the instructor's job.

Under certain circumstances, the instructor can do little, if anything, to inspire in students a readiness to learn. If outside responsibilities, interests, or worries weigh too heavily on their minds, if their schedules are overcrowded, or if their personal problems seem insoluble, students may have little interest in learning. Are you *ready* to learn? Have you reviewed your previous lesson in preparation for the next?

Law of Exercise: This law states that those things most often repeated are best remembered. It is the basis of practice and drill. The human memory is not infallible. The mind can rarely retain, evaluate, and apply new concepts or practices after a single exposure. Students do not learn to weld during one shop period or to perform crosswind landings during one instructional flight. They learn by applying what they have been told and shown (Fig. 3-1). Every time practice occurs, learning continues. The instructor must provide opportunities for students to practice or repeat and must see that this process is directed toward a goal. After solo, *plan* every practice flight, and make every minute count!

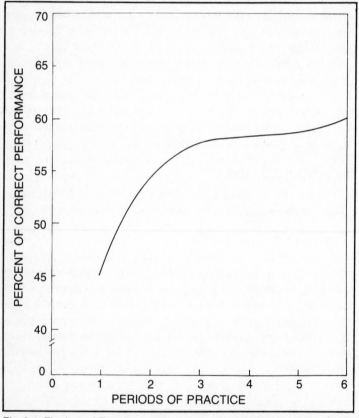

Fig. 3-1. The Law of Exercise is demonstrated by the learning survey. On the average, five to six trials of a new flight maneuver provide about as much learning as the student is capable of absorbing in a single learning session.

Law of Effect: This law is based on the emotional reaction of the learner. It states that learning is strengthened when accompanies by a pleasant or satisfying feeling, and that learning is weakened when associated with an unpleasant feeling. An experience that produces feelings of defeat, frustration, anger, confusion, or futility are unpleasant for the student. Today's instructors understand the benefits of positive (but firm) motivation as a teaching tool. If yours does not, get another instructor.

Law of Primacy: Primacy, the state of being first, often creates a strong, almost unshakable, impression. For the instructor, this means that what is taught must be right the first time. For the student, it means that learning must be right. "Unteaching" is more difficult than teaching. If, for example, a student learns a faulty flight technique, the instructor will have a difficult task in unteaching the bad habit and reteaching a correct one. The first experience must be positive and functional and lay the foundation for all that is to follow.

Law of Intensity: A vivid, dramatic, or exciting learning experience teaches more than a routine or boring experience. A student is likely to gain greater understanding of stalls by performing them than from merely reading about them. The law of intensity, then, implies that a student will learn more from the real thing than from a substitute. In contrast to flight instruction, the classroom imposes limitations on the amount of realism that can be brought into teaching. Mockups, movies, filmstrips, charts, posters, photographs, and other audio-visual aids are used to add vividness to classroom instruction.

Law of Recency: The things most recently learned are best remembered. Conversely, the further a student is removed timewise from a new fact or understanding, the more difficult it is to remember it. It is easy, for example, for a student to recall an airspeed value used a few minutes earlier, but it is usually impossible to remember an unfamiliar one used a week earlier. Instructors recognize the law of recency when they carefully plan a summary for a ground school lesson or a post-flight critique. The instructor repeats, restates, or reemphasizes important matters at the end of a lesson to make sure that the student remembers them. Your reviews of classroom or flight material strengthen learning by applying the law of recency just before your next lesson.

HOW PEOPLE LEARN

Self-concept, how one pictures oneself, is a most powerful determinant in learning. A student's self-image, described in such terms as "confident" and "insecure," has a great influence on the total perceptual process. If a student's experiences tend to support a favorable self-image, the student tends to remain receptive to subsequent experiences. If a learner has negative experiences which tend to contradict self-concept, there is a tendency to reject additional training.

Negative self-concepts inhibit the perceptual processes by introducing psychological barriers which tend to keep the student from perceiving. They may also inhibit the ability to properly implement that which is perceived. That is, they affect unfavorably the "ability to do." Learners who view themselves positively, on the other hand, are less defensive and more ready to "digest" experiences by assimilating all of the instructions and demonstrations offered. It is quite common for a student to be somewhat apprehensive during his or her first few flight lessons, or perhaps a first solo cross-country. With training and experience, this is typically replaced by confidence and an enhanced self-image.

It takes *time and opportunity* to perceive. Learning some things depends on other perceptions which have preceded these learnings, and on the availability of time to sense and relate these new things to the earlier perceptions. Thus, sequence and time are necessary.

A student could probably stall an airplane on the first attempt, regardless of previous experience. Stalls cannot be "learned," however, unless some experience in normal flight has been acquired. Even with such experience, time and practice are needed to relate the new sensations and experiences associated with stalls in order to develop a perception of the stall. Don't expect to master the entire art of flight in your first lesson.

Motivation is probably the dominant force which governs the student's progress and ability to learn. Motivations may be negative or positive, they may be tangible or intangible, they may be very subtle and difficult to identify, or they may be obvious.

Positive motivation is essential to true learning. Positive motivations are provided by the promose or achievement of rewards. These rewards may be personal or social; they may involve financial gain, satisfaction of the self-concept, or public recognition. Negative motivations in the form of reproof and threats should be avoided with all but the most overconfident and impulsive students. Negative motivations are those which may engender fears and be accepted by the students as threats. While they have their uses in certain situations, they are not characteristically as effective in promoting efficient learning as are positive motivations.

Slumps in learning are often due to slumps in motivation. Motivation does not remain at a uniformly high level and may be affected by outside influences, such as physical or mental disturbances or inadequate instructions.

Group approval is a strong motivating force. Every person wants the approval of friends and superiors. Interest can be stimulated and maintained by building on this natural force. Most students enjoy the feeling of belonging to a group and are interested in attaining an accomplishment which will give them prestige among their fellow students. If possible, join some form of flying organization during your training. Keep your motivation high.

LEVELS OF LEARNING

Learning may be accomplished at any of several levels. The lowest level, *rote learning*, is the ability to repeat back something which one has been taught, without understanding or being able to apply what has been learned. Progressively higher levels of learning are *understanding* what has been taught, achieving the *skill to apply* what has been learned and to perform correctly, and associating and *correlating* what has been learned with other things previously learned or subsequently encountered.

For example, a flight instructor may tell a beginning student pilot to enter a turn by banking the airplane with aileron control and applying sufficient rudder in the same direction to prevent slipping and skidding. A student who can repeat this instruction has learned by *rote*. This will never be very useful to the student if there is never an opportunity to make a turn in flight or if the student has no knowledge of the function of airplane controls.

With proper instruction on the effect and use of the flight controls, and experience in their use in straight flight, the student can develop these old and new perceptions into an insight on how to make a turn. At this point, the student has developed an *understanding* of the procedure for turning the airplane in flight. This understanding is basic to effective learning, but may not necessarily enable the student to make a correct turn on the first attempt.

When the student understands the procedure for entering a turn, has had turns demonstrated, and has practiced turn entries until consistency has been achieved in an acceptable performance of those entries, the student has developed the *skill to apply* what has been taught.

The highest level of learning, which is the objective of all instruction, is that level at which the student becomes able to associate an element which has been learned with other segments of learning or accomplishment. The other segments may be items or skills previously learned, or new learning tasks to be undertaken in the future. The student who has achieved this level of learning in turn entries, for example, has developed the *ability to correlate* the elements of turn entries with the performance of complex piloting operations such as those required for the performance of chandelles and lazy eights. The student, with the guidance of the instructor, must achieve this fourth learning level, correlation, to truly master flight.

SAVE MONEY WITH EFFICIENT LEARNING

You can, without doubt, reduce your training costs for either the private or commercial pilot certificate by exercising a few simple practices that take advantage of the laws of learning. The following are examples of techniques which have proven beneficial in both time and money.

Learning Radio Communications

Aviation has a language which is distinctly its own. Full of acronyms, complex terms, and words with special meanings, the new student is faced with a barrier which can be frustrating as well as costly. If the student allows all of his radio communications learning to take place in the crowded and busy confines of a cockpit, the experience will indeed be costly. How then can this block of learning be accomplished more efficiently?

Put the Law of Exercise to work by purchasing a portable radio with an aircraft band and begin the process of listening to aircraft communications at every opportunity at home, in the car, or at the airport terminal. The beginning student should above all adopt this learning technique prior to his first solo! Learn this part of flying on the ground at no cost to yourself, and save that valuable dual cockpit time for learning to fly (Fig. 3-2).

Visit your closest busy airport and simply listen to tower landing, takeoff, and taxi instructions. Then put that learning into practice when you fly. Listen to air traffic control radar communications to learn operations in a TCA or TRSA. Listen to Flight Service Station to learn more about weather communications and flight plans. If available in your area, listen to the TWEB (Transcribed Weather Broadcast). Learn weather by keeping track of what's being reported on the aviation bands. Try listening to 122.8 for

Fig. 3-2. The way to learn air traffic communications is by listening to a portable receiver and watching traffic pattern operations. Learning the language and operations on the ground saves hours of cockpit time.

Unicom communications at non-tower airports or 122.9 for air-to-pilot conversations.

For those located in remote areas, take along your cassette tape recorder the next time you have an opportunity to spend a few hours at a busy airport. Record the various operations and replay the scene in the quiet of your home. Another variation on this theme is to have your flight instructor tape record typical communications sequences, such as an arrival in a TRSA or a departure from a TCA for your listening practice. Listen and learn at no cost to you.

Learning Weather Theory and Practices

The best weather teachers in the world are available to you 24 hours a day, seven days a week, at no cost. Indeed, Flight Service Station personnel are the designated weather briefers for our aviation system. Employ the Law of Intensity by visiting your local FSS. During slack hours, FSS personnel are most willing to teach you weather theory and practices.

Practice reading teletype reports and weather charts just as FSS personnel do. Work up a practice cross-country, then check weather prior to and during your simulated flight. If available, learn to operate pilot self-briefing FSS computer equipment.

There is no such thing as knowing too much about weather. Learn the subject at your every opportunity.

Learning More from Each Dual Flight

Tell me why a student will arrive for an hour of dual *completely* unprepared for learning, with no review of the previous lesson, his thoughts still on business or other parties. Tell me why a student will pay $45 for an hour of dual and not write down a single note on the material covered during the lesson. Tell me why some students begin their first session with a well-organized list of rationalizations as to why their performance is likely to be substandard. Tell me why they have not prepared a well-organized list of questions instead, so that their performance will not be less than adequate. Tell me why some students are completely haphazard in their scheduling of flight lessons, or getting that written passed. Tell me the answer to these questions and I will tell you why some private pilots take from 60 to 80 hours to get their certificate.

The Law of Readiness is real, as are its financial benefits. Arrive at your ground school session with your homework *done*! Arrive at your flight session mentally prepared to *fly*! Forget the rationalizations as to your performance—you are still a student, you are still learning; excellence will come through practice. Take notes on your flying sessions and review them prior to each hour of dual (Fig. 3-3). Plan your solo sessions: list in advance the flight exercises you intend to do. Don't simply bore holes in the sky for hours on end. As a practicing flight instructor, I assure you that the student who arrives prepared and ready to learn will more than receive his money's worth during an hour of dual. Surprisingly, flight instructors are human and find it most discouraging to spend half a session just getting a student prepared to learn and the other half introducing only a part of the new material planned. Want to minimize your training time and still reach an excellent level of proficiency? Invoke the Law of Readiness!

Learning Air Traffic Control

The textbook is not enough. Put the Law of Intensity to work by visiting your local air traffic control facilities. Make an appointment and visit tower, radar, and if possible, traffic control centers. Learn about the other half of the game—the problems that pilots often present to ATC. Understand the role of ATC and your flying will be better and your communications more professional. Again, this kind of learning is at no cost to you. Take advantage of the few things in aviation that are still free.

Learning By Involvement

Make flying a vivid experience in your life. Become involved

Fig. 3-3. You've just paid $40 for a flight lesson. Take time to write down the high points of the session, then review your notes for the next lesson. Mentally re-fly that last hour's exercise. Save dollars by preparation; employ the Law of Readiness to your benefit.

with the art and its practicioners. Where possible, attend FAA or AOPA symposiums, many of which are free of charge. Read supplemental literature on the subject of flying, both lesson-related and historical. If every a subject had a rich and vivid history, that subject is aviation. Put the Law of Intensity to use; save dollars by understanding aviation in depth.

Learning By Practice

The Law of Exercise clearly states that those things most often repeated are best remembered. Before taking your cross countries—

Practice flight computer (E6B) problems ⎫ Private Pilots
Practice reading sectional charts ⎭
Practice taking clearances ⎫
Practice reading approach plates, ⎬ Instrument Pilots
SIDs, STARs and IFR charts ⎭

The cockpit is no place to do homework! Practice on the ground, then apply your learning in the air.

MOTIVATION

To relate an experience in the richness with which it is lived is indeed an impossiblility. One is defeated at the onset, crippled by the inadequacy of the written word. Adjectives fail in their attempt to relay feeling. Long explanations to establish tone are of limited value. To excite drama without vision, without involvement, without witness is an impossibility. The singular tool of writing is thus wholly dependent upon the imagination of the reader for color and embrace. Let us, therefore, begin and do the best we can with the words we have at hand.

Flight instruction is often a miserable occupation. Long hours, low pay, and fast-food meals characterize the profession. The day begins at 7:00 am for the early riser, and ends at 10:00 p.m. with the close of evening ground school. Tonight's class was to be another session of the private pilot ground school program, my 15th time through the same material—nothing new, nothing exciting, just routine. Clean the chalkboard, arrange the visual aids, and await 6:30 p.m., the start time for class.

Each class has its own personality. Some are vibrant, participative, quick to respond and contribute. Some are definitely lackluster. Occasionally a class is downright cold, defiant, continually testing the instructor with clever questions. This class, No. 15, was proving to be of average composure—twelve students of varied backgrounds, ranging from physician to shop foreman to TV repairman to housewife. As the course neared its halfway point, it became apparent that one student, Steve B., was having considerable difficulty with the material. Consistently marginal test grades were received in spite of excellent attendance and absolute attention to every detail of the lectures. Steve B. was stocky, about 195 pounds, age 22, average in appearance and dress, quiet but somehow very intense.

Steve had selected a classroom position immediately to the left of the lectern. I could observe the care with which he took notes and also the tortuous struggle that a test presented. Each question was read several times and each was slowly and methodically answered. At the end of each test or lecture period, I could absolutely count on Steve being among those with further questions to ask.

The low test scores on Steve's part troubled me. How could a person so very intense possible get such low scores? What was I doing wrong—why was I not able to communicate with Steve? Was it apathy on my part? After all, this was the 15th time through the same material. Had my presentations become stereotyped, lacking in enthusiasm?

Following a particularly arduous session on the E6B flight computer, the answer finally came forth. Steve remained after the last questioner had left class and in a quiet way explained his goal: "I want more than anything else to become an airplane mechanic," he said. "But to be a good airplane mechanic, I must know how to fly. That's why I'm here." After a short hesitation, he continued: "But I'm a slow learner—sometimes very slow, like tonight." Our discussion continued and I learned of Steve's background, his previous school experiences, his IQ tests, his learning problem, and also his singular defense. Indeed, life had dealt Steve a short deck. His hand of learning skills lacked the Aces and the Kings; only a pair of Jacks remained to sustain his independent existence—*motivation* and *perseverance*. "I really am learning about flying in this course but it takes me so much longer than the others." This computer—right now I'm terribly confused but I will learn to operate it, *so help me*," continued Steve. Now I began to comprehend the task ahead.

Class No. 15 took on a new meaning. With Steve present, it developed its own unique personality. I had seen motivation and learning go hand in hand before but never, absolutely *never* to the extent put forth by Steve. Things became more relaxed in class. I directed every important lecture point at Steve, perhaps at the expense of the other students. After-class question-and-answer periods with Steve were approached with a new awareness of need. No, there wasn't any real change in Steve's continuing low grades but there now was *understanding*. In a class of the average and above, Steve was somehow keeping at a passing 70% level—a major accomplishment for a slow learner in a fast-moving course.

It was several weeks after the completion of the course before I again saw Steve. Preoccupied in preparing a lesson plan, I hardly noticed when a hand quickly placed an FAA private pilot written exam test report over the top of my work papers. Only two words were spoken: "I passed." Steve had passed; the score was only 72, but Steve had passed! I was elated, overjoyed! I grabbed Steve's hand and shook it vigorously. After hearty congratulations, we talked briefly about Steve's plans to become a mechanic. A major obstacle had been passed by sheer motivation and perseverance. Next came mastery of flight itself in conjunction with mechanic's training. Yes, it will be a slow process for Steve but he will make it. In the final analysis, it's motivation that counts, the kind that I saw Steve put forth. Class No. 15 had an unforgettable personality. If a Steve can overcome a true learning disability, any student of average or above learning capabilities can, indeed, master the art of flight—if motivated!

FAA PUBLICATIONS AND TRAINING MATERIALS

The following gives sources for supplemental study materials. Many of the documents offered as such are available at little or no cost. In particular, the Advisory Circulars are recommended.

Checklist

Advisory Circular 00-2, *The Advisory Circular Checklist and Status of FARs*, contains a list of current FAA advisory circulars and Federal Aviation regulations, together with their status as of a given date, contents, and cost. The checklist is updated triannually and provides detailed instructions on how to obtain both advisory circulars and Federal Aviation Regulations. It also contains a list of GPO bookstores located throughout the United States which stock many government publications. The checklist may be obtained free upon request from the U.S. Department of Transportation, Publications Section, TAD 443.1, Washington, D.C. 20590.

Federal Aviation Regulations

The following FAR Parts are those you may be most interested in reading. They pertain primarily to the operation and maintenance of the aircraft and to obtaining a pilot's certificate or an airframe and powerplant mechanic certificate. Parts 1, 61, and 91 are required reading for either a private pilot or commercial pilot certificate. In addition, NTSB regulation Part 830 is also required.

☐ Part 1 Definitions and Abbreviations.
☐ Part 21 Certification Procedures for Products and Parts.
☐ Part 23 Airworthiness Standards: Normal, Utility and Acrobatic Category Aircraft.
☐ Part 33 Airworthiness Standards: Aircraft Engines.
☐ Part 35 Airworthiness Standards: Propellers.
☐ Part 39 Airworthiness Directives.
☐ Part 43 Maintenance, Preventive Maintenance, Rebuilding, and Alteration.
☐ Part 61 Certification: Pilots and Flight Instructors.
☐ Part 65 Certification: Airmen Other Than Flight Crewmembers.
☐ Part 91 General Operating and Flight Rules.

The FARs may be purchased from the Superintendent of Documents, U.S. Government Printing Office, Washington, D.C. 20402. You should include a check or money order payable to the Superintendent of Documents with each order. Refer to Advisory Circular 00-2 for the correct pricing and ordering information.

Advisory Circulars

Advisory circulars are issued by the FAA to inform the aviation public, in a systematic way, of nonregulatory material of interest. The contents of advisory circulars are not binding on the public unless incorporated into a regulation by reference.

Request free advisory circulars from: U.S. Department of Transportation, Publications Section, TAD 443.1, Washington, D.C. 20590. Persons who want to be placed on the FAA's mailing list for future circulars should write to the U.S. Department of Transportation, Distribution Requirements Section, TAD 482.3, Washington, D.C. 20590. Be sure to identify the subject matter desired, as separate mailing lists are maintained for each advisory circular subject series.

Order "for sale" advisory circulars from: Superintendent of

Documents, U.S. Government Printing Office, Washington, D.C. 20402, or from any of the GPO bookstores located throughout the United States. Use AC 00-2, Advisory Circular Checklist, for the cost of each circular.

Notices of Proposed Rule Making (NPRMs)

The Government Printing Office will send NPRMs to all purchasers of the Federal Aviation Regulation containing the particular regulation concerned.

Films

FAA—Order free catalog from: FAA Film Library, AAC-44.5, FAA Aeronautical Center, P.O. Box 25082, Oklahoma City, Oklahoma 73125.

U.S. Air Force

For information about Air Force pilot training manuals, write: USAF Central Audio-Visual Library, AF Audio-Visual Center, Norton AFB, California, 92409.

Aviation Safety Statistics and Reports

For safety information, write: National Transportation Safety Board, 800 Independence Avenue, Washington, D.C., 20594.

Chapter 4
Flight Training

Flight training begins on the ground, in the classroom. To be dollarwise efficient, the first step is ground school, regardless of the rating sought—private, instrument, or commercial. The fundamental purpose of ground school is to prepare a student for his flight training activities. In that light, it is essential that ground school training keep *ahead of* flight knowledge requirements. Federal Aviation Regulations and airport operations must be understood by a student prior to solo; navigation fundamentals must be mastered prior to cross-country flight; approach plates and en route charts must be comprehended before instrument flight. The learning lead is ground school—flight follows.

The manner in which ground school and flight training is integrated is essentially fixed for college-level flight programs but is left to the discretion of the student for FBO or freelance flight instruction programs. The two most favorable options are to complete the ground school program (including the written) before beginning the flight program, or to integrate the ground school program with the flight program, keeping the ground training program well *ahead of* the flight program.

Of the two approaches, the first is preferable for the beginning private pilot. By so doing, you will have had an overview of the business of flight before spending dollars on flight training. In short, by completing a private pilot ground school as the first formal step in training, you can then decide whether or not to continue flight training. Should the answer be negative, the only cost becomes that of the ground school, a very minimal expenditure indeed. Assuming that you are going to continue, you are now in the favorable position of having ground school training completed. With the necessary background of aeronautical knowledge established, you can then concentrate entirely on the flight training portion of the program. The large dollar sums are spent on flight training. Anything that will improve efficiency during this part of your training is, indeed, a significant dollar-saver.

INTRODUCTION TO FLIGHT TRAINING

By way of introducing the details of the private pilot and instrument/commercial pilot training programs, an overview is necessary to illustrate the general method by which training progresses. By understanding the principal building blocks and their relationships one to the other, the program takes on a more meaningful theme. Unfortunately, many beginning students have not been shown the threads of continuity so necessary to their training. They are left with nagging questions: "Why must I learn all about stalls? What are they good for? Why must I do turns about a point? What do S-turns across a road prove? I came here to learn to fly, not to do a lot of expensive maneuvers that I never intend to do again." Let us try to answer some of these questions in the next several paragraphs.

Private Pilot Training, an Overview

Given the fact that you have taken your Class III medical and have sent for an FCC Restricted Radio-Telephone license, training begins by establishing two foundation stones. The first is a general background of aeronautical knowledge (ground school) and the second is basic flight technique. Approximately one-third of the flight program is devoted to basic flight technique. It is during this effort that you learn the whys about stalls and ground reference maneuvers and the fact that these are necessary preliminary steps in learning to land. This vital portion of the program is where you really learn to fly the aircraft.

Having mastered these two foundation stones, the next major event is solo flight. You are given the opportunity to practice basic flight maneuvers (except for emergency landings) on your own. The halfway point has been reached.

The remaining half of the program consists of several dual and solo cross-country flights. This is by far the most pleasurable part of the effort. Going somewhere (at last) puts to use all that was learned in ground school and basic flight technique practice.

In preparation for the FAA flight test, the instructor will spend at least three hours going over training high points and polishing flight technique. When all is satisfactory, he will review the results of your FAA written examination to correct any deficiencies noted (questions missed) and then provide a recommendation for the flight test. The next and final step is simply demonstrating to the FAA examiner that you, indeed, have mastered the art cf VFR flight (Fig. 4-1).

Instrument/Commercial Pilot Training, an Overview

Training relating to the instrument rating or the commercial pilot certificate proceeds along a route very similar to that of the private pilot. Again, two basic foundation stones are to be put in place—instrument/commercial ground school and basic instrument flight technique. At least ten hours will be devoted to basic instrument flight technique. Climbs, descents, turns, slow flight, stalls, recovery from unusual attitudes, and the like are to be mastered during beginning training. It is during this phase that the student who is bothered with vertigo learns to overcome the problem.

The results of basic instrument flight technique training and instrument ground school blend together in the next program phase, learning instrument approaches and holding patterns. These ma-

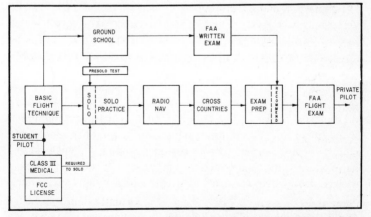

Fig. 4-1. Private pilot training is broken into phases beginning with basic flying technique training and ending with cross-country and examination preparation training. To solo, a student must have his or her Class III medical, an FCC Radio-telephone permit, and a flight instructor sign off on his or her student permit and logbook.

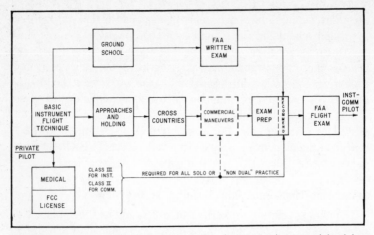

Fig. 4-2. In a manner similar to the private pilot, instrument /commercial training begins with basic IFR flight technique training. To be of maximum benefit to the student, cross-country training must be accomplished for the most part in a busy terminal environment. Simply flying long straight IFR routes builds hours but adds little to one's proficiency.

neuvers require discipline and precision, hence necessitate considerable practice. At the end of this phase the halfway point has been reached.

Once again, the more pleasurable phase of the program is reached, the IFR cross-country flights. If the commercial certificate is being pursued, VFR maneuvers and VFR cross-country flights will be integrated into the program at points to complement instrument training. In reality, about 70% of the commercial certificate is the instrument rating (Fig. 4-2).

Just as for the private certificate, the flight instructor is required to spend time at the end of the course preparing you for the FAA flight examination. After he is assured of flight proficiency and has critiqued your written examination test score, his recommendation states that you are prepared. Granted, the instrument flight test (or instrument portion of the commercial) is one of the most difficult. However, your instructor's recommendation will not be given unless you are fully capable of passing the test. His recommendation signifies confidence; approach your test with similar conviction.

The Private Pilot Written Examination

As a part of the effort in preparing for a private pilot's examination, you must pass a written test on the theory of flight, meterology, the Airman's Information Manual, aviation regulations, use of the flight computer, navigation, preflight planning, radio procedures, emergency procedures, and basic attitude flying, using only aircraft instruments for reference. The minimum passing grade for the written test is 70. In general, the written is given at a local FAA facility and requires typically three hours or so to complete. You need supply only a navigation computer and plotter; textbooks or notes of any kind are forbidden. An electronic calculator is permitted. However, it must be battery-operated, and contain no formulas. The written test is valid for a period of 24 months. In the event you fail a written test, you may apply for retesting 30 days after the date you have failed the test or upon presenting a statement from an instructor stating he has given you additional instruction and considers you competent for retesting. Although ground school is not currently required by regulation, many students find this an efficient means of preparing for the written examination. As a minimum, the following study materials are recommended in preparation for the private pilot written examination:

☐ *Pilot's Handbook of Aeronautical Knowledge*, AC 61-23B.

☐ *Aviation Weather*, AC 00-6A.

☐ *Aviation Weather Services*, AC 00-45B.

☐ Federal Aviation Regulations, Parts 1, 61, and 91, and National Transportation Safety Board Investigation Regulation, Part 830.

☐ *Federal Aviation Regulations Written Test Guide*, AC 61-34A.

☐ *Private Pilot-Airplane, Written Test Guide*, AC 61-32C.

☐ *Private Pilot-Airplane, Flight Test Guide*, AC 61-54A. (Included in this text as a part of Chapter 8.)

The Instrument Written Examination

Without doubt, the instrument written examination is the most difficult FAA test other than the Airline Transport Pilot examination. The test is detailed in that it demands intimate knowledge of IFR regulations and air traffic system procedures, methods of IFR navigation (VOR, ADF, ILS, RNAV), the proper use of IFR charts and approach plates, and in-depth understanding of weather data. The latter includes weather reports and forecasts and the development of forecasts based on trends. In preparation for the written, it is absolutely essential that you be able to read well the four basic TWX reports (Surface Aviation, Forecast Area, Forecast Terminal, and Forecast Winds Aloft) and the four basic weather charts (Surface Analysis, Weather Depiction, Prognostic, and Radar).

Allow at least four hours for the test. The minimum passing grade is 70. The test is good for 24 months. In addition to the Federal Aviation Regulations and weather books noted as reading for the Private Pilot written examination, the following are highly recommended for reading in preparation for the instrument examination:

☐ *Instrument Flying Handbook*, AC 61-27B.
☐ *Instrument Rating Written Test Guide*, AC 61-8D.
☐ *Passing Your Instrument Pilot's Written Exam*, TAB Book # 2255.

The Commercial Written Examination

The commercial written examination dwells heavily on knowledge of aerodynamics and airplane operations, particularly as associated with complex aircraft. The test contains questions relating to the use of flaps, retractable landing gears, controllable propellers, high-altitude operation (with and without pressurization), and airplane performance charts and tables. The latter includes weight and balance computations as may apply to a light twin-engine aircraft. In addition, the test covers Federal Aviation Regulations and ATC procedures as they pertain to commercial operations and the commercial pilot.

While the test is not difficult, it does require in-depth preparation. As with other FAA written examinations, the minimum passing grade is 70. The test is good for 24 months. The following are highly recommended as reading in preparation for the commercial written:

☐ *Flight Training Handbook*, AC 61-21A.
☐ *Commercial Pilot Written Test Guide*, AC 61-71B.

THE STUDENT PILOT

To obtain a student pilot certificate, you must be at least 16 years of age, able to read, speak, and understand the English language, and qualify for at least a third-class medical certificate. A combination medical certificate and student pilot certificate will be issued, at your request, by the medical examiner upon the satisfactory completion of your physical examination. Student pilot certificates may be issued by FAA inspectors or designated pilot examiners if you already possess a valid medical certificate. Since a medical is required prior to solo flight, it is really simplest to obtain a student pilot certificate by visiting your aviation medical examiner as the initial step in learning to fly. A combination medical and student pilot certificate is valid for a period of 24 calendar months. The certificate expires at the end of the calendar month in which it was issued two years hence. For example, a certificate issued on January 2, 1983, will expire at midnight on January 31, 1985.

To solo is to taste adventure. A high point in any pilot's career, the initial solo requires adequate preparation. Regulations specify that, as a minimum, a student pilot be given training in aircraft preflight inspection, operating the aircraft engine, taxiing, takeoff, landing, traffic pattern procedures, level flight, turns, climbs, glides, stalls, and emergency landings prior to solo. Upon completion of these requirements and demonstration of competence in the control of an aircraft, a flight instructor may endorse your certificate that you are competent to solo. This endorsement must be made prior to the first solo flight and is an endorsement to solo *only in the make and model of aircraft so designated* by the flight instructor. A solo endorsement is also required in your logbook. Except for special cases, this solo logbook endorsement is valid for a period of 90 days. After this time, you must fly with and have your logbook endorsed again by a flight instructor in order to continue solo flight. These endorsements permit student solo only in a local area designated by the flight instructor. When you are endorsed to solo in one type aircraft, you may not solo in another unless your flight instructor has endorsed your student pilot certificate and logbook to permit soloing in both aircraft types.

Solo cross-country flight is perhaps the high point in student training. Regulations require that, prior to solo cross-country flight, you be given instruction (and demonstrate competence) in basic flight planning elements such as plotting courses, evaluation of weather reports, estimating time en route, and fuel required. In addition, competence in crosswind and simulated soft-field takeoffs and landings, climbing and gliding turns at minimum safe airspeeds, and cross-country navigation by reference to aeronautical charts is required. Knowledge of safe operating procedures in simulated emergencies such as engine failure, loss of flying speed, marginal visibility, deteriorating weather, becoming lost, and other critical situations is necessary. Skill in conforming with air traffic control instructions by radio and lights, the proper use of two-way radio communications, VFR navigational procedures, and simple maneuvers by reference only to instruments is required prior to solo cross-country flight. When you demonstrate competence in the required skills, a flight instructor may endorse your student certificate for solo cross-country work. A logbook cross-country endorsement must be made by a certificated flight instructor prior to *each* solo cross-country flight. You must carry your logbook on each solo cross-country flight. Your student pilot certificate must also be carried on all solo flights. In the event operation of an aircraft radio transmitter is anticipated, you must obtain an FCC radio telephone operator's permit. This permit can be obtained by sending FCC Form 753-A to the Federal Communications Commission. These forms are usually available at flying schools. Once obtained, a radio telephone operator's permit is currently valid for life.

Federal aviation regulations prohibit a student pilot from carrying passengers, or operating an aircraft for compensation or hire, or in the furtherance of a business. In addition, a student pilot is not permitted to make international flights. The purpose of the student pilot certificate is purely for training; extended flight privileges are reserved for those who pass the private pilot examination (or commercial, etc.).

PRIVATE PILOT CERTIFICATE

To be eligible for a private pilot certificate, you must be at least 17 years of age, able to read, speak and understand the English language, hold at least a third-class medical certificate, and within 24 months have passed a written examination for the rating of private pilot. In addition, you must have at least 40 hours of flight instruction of which 20 hours are solo time and 3 are night flight. Of the solo time, at least 10 hours must be devoted to cross-country flight.* One cross-country flight must include three landings at a place more than 100 miles from the point of departure.

Following your first solo cross-country, regulations require that your flight instructor provide at least three hours of training to include maneuvers previously learned, as well as additional maneuvers required for the private pilot test. While the regulations specify a minimum of 40 hours total flight time (35 hours for a certificated flying school), a typical value for the average student is 53 hours. With the increasing complexity of flight, we can expect this to approach 60 hours in the near future. The reason is simple: Regulations specify minimum hours only from the standpoint of aeronautical experience. *The real requirement is proficiency, not hours!*

---

*A complete listing of hours requirements will be found at the end of this chapter.

When you have acquired the necessary qualifications and have obtained a written recommendation for a flight test from an appropriately rated flight instructor, you may apply for the private pilot flight test. The flight test can be taken at any FAA General Aviation District Office or from a designated pilot examiner. There is no charge for flight tests when conducted by an FAA inspector; however, designated pilot examiners are entitled to charge a reasonable fee.

In actual practice, you will find that the FAA does not have staffing adequate to conduct private pilot or commercial pilot examinations. Hence, these are usually performed by designated pilot examiners. A list of examiners for your area can be obtained from the FAA.

A private pilot test consists of three phases: an oral examination, a basic piloting technique, and a cross-country test. Regulations require that you perform the following procedures and maneuvers:

1. *Phase I—Oral Operational Test:* You will be required to present and explain aircraft registration, airworthiness and equipment documentation, as well as airplane logbooks and airworthiness inspection reports. To determine that you know what performance and operating information is important, you are required to demonstrate a practical knowledge of aircraft performance parameters, range, operation, weight and balance, etc. In addition, the oral test covers aircraft preflight procedures and the use of radio for voice communications.

2. *Phase II—Basic Piloting Technique Test:* The objective of Phase II is to evaluate your flying skill. A demonstration of preflight procedures, taxiing, normal and cross-wind takeoffs and landings, climbs, level flight, descents, and flight at minimum controllable speeds is required. Stalls and stall recovery, 720-degree steep turns about a point, full stall landings, short and soft field takeoffs and landings, and engine-out emergency landings are required.

3. *Phase III—Cross-Country Flight Test:* The objective of this phase of testing is to determine whether you can effectively prepare for a cross-country flight in a reasonable period of time and furthermore conduct such a flight in a safe and expeditious manner using normally available aids and facilities. Before takeoff for the flight test, you will be requested to plan a cross-country flight to a point at least two hours cruising range distance in the airplane to be used for the test. At least one intermediate stop will normally be included. You are expected to procure pertinent available weather informa-

tion, plot the assigned course, establish checkpoints, and estimate flying time and fuel requirements. The use of the *Airman's Information Manual* for reference information and a flight computer for dead reckoning computations is expected.

As a part of the flight test, the examiner will request that you demonstrate cross-country flying ability by following a designated course, the use of radio aids in VFR navigation, and instrument flight. During simulated instrument flight, you must be able to recover from the start of a power-on spiral, recover from the approach to a climbing stall, and execute normal turns of at least 180 degrees to within plus or minus 10 degrees of a preselected heading. Shallow climbing turns to a predetermined altitude, shallow descending turns at reduced power to a predetermined altitude, and straight and level instrument flight are required.

During a flight test, the examiner acts in the capacity of an observer; you are pilot-in-command. The examiner simply poses the problem; you must demonstrate the correct response. A successful flight test is denoted by obvious mastery of the aircraft; the outcome of a maneuver must never be in doubt. If you fail a flight test, you may apply for a retest upon presenting a statement from your flight instructor that you have been given additional instruction and the flight instructor now considers you ready for retesting.

Having passed the private pilot examination, it is generally desirable to transition into a high-performance aircraft (an aircraft that has over 200 horsepower or that has retractable gear, flaps, and a controllable propeller.) To do so requires flight instruction from a CFI who is rated to fly the aircraft involved and a logbook CFI sign-off certifying that the student is competent to operate a high-performance aircraft. Many instructors believe that a new private pilot should have at least 200 hours before making this transition.

THE INSTRUMENT RATING

To be eligible for an instrument rating, you must hold a current private or commercial* pilot certificate, read and speak English, and have passed the instrument written examination within the past 24 months. You must be knowledgeable in subjects related to IFR flight, such as IFR air traffic procedures and regulations, IFR navigation by VOR, ADF, ILS and RNAV, IFR charts, and weather theory and forecasting.

---

*Commercial in this context refers to the limited commercial certificate for which an instrument rating is not required.

Flight experience required, as a minimum, must total at least 200 hours of which at least 100 are pilot-in-command hours and 50 are cross-country hours. In addition, 40 hours of IFR training is required.**

The instrument rating test consists of an oral examination followed by a comprehensive evaluation flight. During the flight, you must navigate in an ATC environment (simulated or real), accept clearances, respond to ATC vector instructions, conduct approaches by means of VOR, ADF, and ILS, and operate the aircraft under emergency circumstances (equipment failures and unusual attitudes).

A FEW IFR THOUGHTS

The instrument rating is demanding, confounding, intricate, sometimes vexing, an ultimate challenge—unquestionably, a difficult rating to achieve so filled with procedures and charts and endless kinds of limits, MEAs, DHs, MOCAs, MDAs . . . and on and on.

Once vision of our outside environment is removed, we are forced into secondary procedures—flight by reference to instruments. Denied our single most important data input—external vision—flight takes on a whole new technique. No *single* instrument tells the *whole* truth. Data bits and measurements must be laboriously extracted from each of the many panel dials to establish a flight attitude. No single chart denotes the route. Information must be assembled from SIDs and STARs and en route charts and approach plates and . . . the list goes on and on. There's no room for error; when approaching a decision height of 200 ft AGL, you simply do not make a mistake in reading the altimeter. Second chances are virtually unheard of. Demanding of discipline and unforgiving of miscalculation, IFR flight simply requires training appropriate to the task. It can be no other way!

There are pilots and there are instrument pilots. Only when you have achieved your instrument rating have you truly joined the pilot fraternity. All who hold the rating have made a sacrifice of time and dollar to obtain this rating. There are no shortcuts to the process; the ultimate task to be accomplished simply does not allow them. You, too, must dedicate yourself to the challenge if indeed you desire an instrument rating. An ominous task? No, indeed. The thrill of learning something new is never ominous. Also, overcoming a

---

**A complete listing of hours requirements will be found at the end of this chapter.

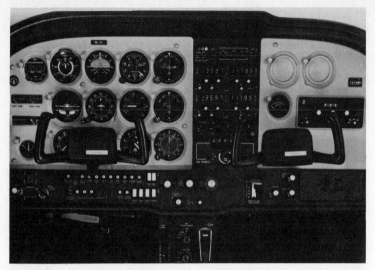

Fig. 4-3. The Cessna 172 Skyhawk instrument panel provides a full IFR capability with dual 720 Nav-Comms, ADF, transponder, and single-axis autopilot.

truly difficult challenge results in a worthwhile sense of achievement, something to be remembered for a lifetime.

What then are the facts? What shall you plan on? How can you achieve the most learning for the dollar? To begin, recognize that the instrument program takes time; there are many new subjects to be mastered. Estimate about eight months of time required unless you are a full-time student at an aviation college. Secondly, recognize that the 40-hour training minimum established by the FAR's is seldom met. A figure of 50 hours for the dedicated student and 60 hours for the average student is more realistic. *Absolutely* attend a high-quality formal ground school; seek out the best in your area and pay the extra dollar, if necessary. The nuances of the ATC system simply are not in textbooks! They must be learned from an experienced instructor.

Next, insist that your training aircraft be fully equipped with modern, properly operating, calibrated communications-navigation avionics. The ADF, transponder, altitude encoder, and DME are simply a part of modern day instrument flight (Fig. 4-3). Training with these items is a must. Do not allow gaps in your training by flying obsolete or poorly calibrated avionics. Rent only aircraft that are equipped and capable of doing the job of providing IFR training. Stated perhaps a bit harshly, but very factually, the day of the 90-channel shared single receiver nav-comm and the "coffee grinder" ADF is over. Such equipment is not capable of meeting today's demands.

Finally, beware of the procedural trainer (simulator). The purpose of the ground instrument simulator is simply to *introduce* instrument procedures—nothing more! Currently, the FARs will accept up to 20 hours of instrument training credit from a simulator. Unless you have access to a *very* high simulator, minimize your simulator hours in favor of real ones in the airplane. Use the simulator only for its intended purpose—the learning and initial practice of procedures. Do not build your instrument time while seated comfortably in a chair driving a trainer that can't move, stall, impose G forces, or induce apprehension. To the extent possible, build your instrument time in the cockpit, where the environment is real. Fly at night as often as possible. You are paying good dollars for your rating—take your training in as realistic as an environment as possible. A good "India ink" night is hard to beat for realism! In short, beware the simulator; some are hardly more than toys. If your trainer doesn't accurately model a complete aircraft cockpit, you are wasting your money.

COMMERCIAL PILOT CERTIFICATE

To qualify for a commercial pilot certificate, you must be at least 18 years of age, be able to read, speak, and understand the English language, hold a valid first or second-class medical certificate, and have the aeronautical experience required. This includes an instrument rating which you must hold or concurrently obtain as a part of your commercial flight test. In addition, to take the commercial flight test, you must have passed within 24 months the commercial pilot written examination and have a written recommendation from a flight instructor stating that you have qualified for a commercial flight test. An instructor's recommendation for a flight test is valid for 60 days.

Aeronautical experience required for the commercial certificate consists of a minimum of 250 hours of total flight time, including at least 50 hours of flight time in powered aircraft. In addition, 100 hours of flight time must be as pilot-in-command, including a minimum of 50 hours of cross-country time.* Cross-country experience must include takeoffs and landings from different airports under two-way radio instruction from an airport tower. One cross-country flight must be accomplished to include three landings, each

---

*A complete listing of hour requirements will be found at the end of this chapter.

of which is at least 200 nautical miles from the other two points. In preparation for the commercial flight test, you must have a minimum of 10 hours of instruction in a complex aircraft and 10 hours of flight instruction devoted to maneuvers required for the commercial pilot flight test. The latter is usually given in a complex aircraft to conform to the regulation which states that an applicant must have 10 hours of flight instruction and practice in an airplane having retractable landing gear, flaps, and a controllable propeller. To qualify for night flight under ICAO (International Civil Aviation Organization) requirements, you must have at least 5 hours of flight time at night, including at least ten takeoffs and ten landings as pilot in command.

The commercial pilot flight test is conducted in four phases: an oral examination, basic flying techniques, precision flight maneuvers, and a cross-country flight. The test is similar in principle to the private pilot examination, except that a greater depth of knowledge, accuracy, and flying skill is expected. In addition, you must demonstrate the required precision flight maneuvers. These maneuvers consist of gliding spirals, lazy eights, steep turns, chandelles, maneuvering at minimum controllable air speed, and stalls from all normally anticipated flight attitudes with and without power. Accuracy landing within 200 feet beyond a designated mark are also required.

ADVANCING THROUGH THE RATINGS

For business reasons or simply to add to their flying proficiency, a sizable number of private pilots decide every year to advance through the ratings. The most commonly sought rating is the instrument, followed by the commercial. For those who have decided to follow aviation as a career, the flight instructor and multi-engine ratings are the next two rungs in the ladder; the Airline Transport Pilot rating is the top step of achievement. Considering fixed-wing land airplane pilot ratings, the path to the top is as shown in Fig. 4-4.

Other related ratings include rotorcraft, glider category certificates, and single-engine seaplane, multi-engine seaplane, gyroplane, helicopter and free balloon class ratings. Ground instructor ratings are also available.

In this text our concern is with the basic group, i.e., through the commercial rating. Table 4-1 synopses applicable hour requirements.

To read the hour values shown in Table 4-1, note that the rows with subcategories break down the total into their major parts. For example, the total flight time of 40 hours for the private pilot *includes* 20 hours of VFR dual which *includes* 3 hours of night flight, 3 hours of cross-country flight, and 3 hours in preparation for the flight test. In addition, the 40 hours includes 20 hours of VFR solo flight which includes 10 hours in an airplane and 10 hours of cross-country flight. Reading across the table, the hours are also inclusive. For example, the 50 hours of required flight instruction time for the commercial rating can include all instruction time accrued for the private and the instrument ratings. Let us examine the following example:

Jim R. has his private pilot certificate and his instrument rating. He has accrued 300 hours of total time of which 10 is night solo (pilot-in-command) and 200 is VFR day flight as pilot-in-command. In addition, Jim has 100 hours of cross-country time as pilot-in-command, of which 30 are in a complex aircraft. If Jim were to pursue his commercial certificate, he needs only to: pass the commercial written, spend 10 hours in preparation for the commercial flight test, and pass the commercial flight test.

As a note of clarification relating to the instrument rating, a total of 40-hours of *instrument flight instruction* is required. Of

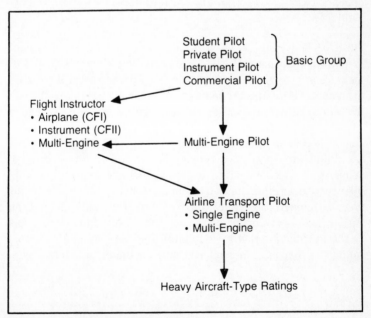

Fig. 4-4. Progression through the ratings.

Table 4-1. FAR Flight Time Requirements.

Time Category		Private Pilot Certificate	Instrument Rating	Commercial Pilot Certificate
EXPERIENCE				
1. Total Flight Time (all)			200	250*
2. Total PIC** Time			100	100
(a) Night				5
(b) Cross Country			50	50
(C) Airplane				50
TRAINING				
3. Total Flight Instruction Time		40	40	50
(a) VFR Dual		20		
• Night		3		
• Cross Country		3		
• Test Preparation		3		10
• Complex Aircraft				10
(b) VFR Solo		20		
• Airplane		10		
• Cross Country		10		
(c) Instrument			40	10
• Simulator			20	
• Dual (CFII)			15	
• Airplane			5	5

• Cannot include more than 50 hours of instruction time
** Pilot-In-Command

these, 15 hours of *instruction* must be given by a Certified Flight Instructor-Instrument (CFII). The remaining 25 hours of instrument *training* can be given by a Certified Flight Instructor (CFI). A maximum of 20 hours of training can be accomplished in a simulator. In reality, there is *no* requirement for simulator hours as such; all of the required 40 hours of instrument instruction time can be accomplished in an airplane.

Pilots are aggressive people. The desire to improve flying competence is commonly seen throughout the association. One pilot out of every three is instrument rated and one out of every four holds a commercial certificate. One out of every thirteen holds an Airline Transport Certificate and one out of every fifteen pilots is a Certified Flight Instructor. Your future, too, may well see additional ratings as your pilot-in-command time continues to build.

Chapter 5
Buying an Aircraft

If you've decided to buy an aircraft, congratulations! Perhaps the decision is not quite as important as the decision to get married, buy a home, or start a college degree program, but it certainly ranks a close second. There are only 210,000 general aviation airplanes in the United States. Considering the fact that many of these are owned by commercial operators or corporations, individual aircraft ownership is still a remarkable event. Regardless of the reason—business, training, or recreation—aircraft ownership carries with it a sense of pride. To fly one's own plane is indeed a feeling of fulfillment. To fly *where* you want *when* you want is freedom in one of its purest forms.

Now that you have made your decision, let's look at the purchase process. For the new aircraft, the matter is quite simple. The primary consideration is generally cost, followed closely by the complement of radio and navigation equipment desired by the purchaser. The used aircraft, however, requires considerably more attention. Since a used aircraft is often the choice of the first-time buyer, the next few paragraphs have been included as an aid to the somewhat more involved problem of purchasing a reliable, used aircraft at a reasonable price.

PURCHASING A USED AIRCRAFT

Once the decision is made to purchase an aircraft, it often results in "buying fever." A strange malady, buying fever causes a pilot to haunt aircraft showrooms, read *Trade-A-Plane* at breakfast, and purchase flying magazines just for the classified ads. The race is on—*buy now, buy now,* sings that inner voice. Suddenly *all* airplanes start looking good, even ultralights. Taking the family out for a drive, naturally it's to the airport—where else? Making vacation plans? "Well, let's see, where can we fly to? Hmm, I seem to remember that Jim owned his own aircraft—haven't talked to Jim in ten years—guess it's about time I give him a call."

Buying fever is a delightful affliction which quickens the pulse and gives a host of new personal plans for the brain to relish. Out with thoughts of work; away with distressing problems. It's time to buy an aircraft, to fly—away, far away. Bermuda? Glacier National Park? Jackson-Hole? Fishing off the coast of Baja? Hunting in Canada? Golf at Palm Beach? The thoughts continue to roll on as the temperature of buying fever mounts. A cure? None known at present. The malady is terminal—an aircraft *will be purchased*. Recognizing that the end is inevitable, the following is offered as a checklist in the hope that it will provide some guidance to your purchase, even in the most vicious throes of the buying-fever syndrome.

The High-Time Aircraft

The cost of a used aircraft varies significantly with age and the number of hours on the engine. The number of hours on the airframe is not too significant as long as it has had reasonable care. An example of this is the used flying school aircraft—many hours to be sure, but given regular maintenance, an airframe with 3000 hours or so is not an unreasonable buy, provided the engine has less than 1000 hours on it, preferably less than 500. The value of an aircraft tends to oscillate with engine hours, being highest at zero engine time and lowest when the engine reaches TBO (Time Between Overhaul). For a single-engine aircraft, the TBO is typically 1500 to 2000 hours. For some twins, the engine TBO may be only 1200 hours.

As a first step in evaluating a tentative selection, take a close look at the aircraft maintenance records. A well-kept set of maintenance records, which properly identify all previously performed maintenance, alterations, and AD compliances, is generally a good indicator of the aircraft condition. This is not always the case, but in any event, before you buy, require the owner to produce the maintenance records for your examination, and require correction of any discrepancies found on the aircraft or in the records. Many prospective owners have found it advantageous to have a reliable, unbiased maintenance person examine the maintenance records and the aircraft before negotiations have progressed too far. The small cost for a review by a qualified mechanic is your best insurance against the purchase of a "dog."

As a second step, have a pilot who has time in the type of aircraft being considered give the machine a test flight. Get his evaluation of the avionics equipment, navigation instruments, engine performance, etc. Know in advance what you are buying! There is nothing more disappointing than buying an aircraft with hard-earned cash only to find out that the gyros must be overhauled, or the radio can't be repaired, or the retractable landing gear hydraulic cylinders must be overhauled. Does this happen? Regrettably, yes; buying fever can take its toll *if you let it!* Let's take a look at just a few actual cases:

☐ Aircraft purchased, nav-comm inoperative. Upon examination, it was discovered that the nav-comm had been gutted for spare parts; only a shell remained. Cost—$600 for a used replacement.

☐ Aircraft purchased, minor fuel seepage around gasoline tanks noted. Cost—$1500 for new fuel cells.

☐ Aircraft purchased, low static rpm and poor climb noted. Cost—$4000 for engine overhaul, engine logbook "unclear" as to actual engine time.

☐ Aircraft purchased, subsequent annual inspection disclosed that the aircraft was fitted with the wrong propeller. Cost—$1000 for a new propeller.

☐ Aircraft purchased, autopilot inoperative. Examination disclosed improper gyros installed in aircraft to mate with autopilot. Cost—$700 to repair.

☐ Aircraft purchased, battery low. Cost—$160 for a new 24V battery.

☐ Aircraft purchased, engine magneto AD not complied with. Cost—two new magnetos.

And so it goes, on and on. The list has a common theme. First, the aircraft was purchased and *later* it was examined—buying fever strikes again. Let's try reversing the process. There is nothing wrong with buying an aircraft with a deficiency so long as the extent of the deficiency is clearly known to the buyer and the cost of the aircraft adjusted accordingly. As an absolute minimum, have *at least* the following items checked *before* signing a purchase contract:

   1. Engine:
    a. Thorough engine logbook review.

b. Compression check.

c. Spark plug check (for evidence of oil consumption, etc.).

d. Magneto and wiring harness examination.

e. Check for oil seepage (leaking gaskets, worn magneto seals).

f. Exhaust manifold and heater check for leaks which might introduce carbon monoxide into the cabin.

g. Check general condition of hoses, alternator belt, etc.

h. If engine has been idle for three months or more, check cylinder walls for rust (boroscope examination).

i. Engine type history—check for a history of known engine design faults. Certain engines with light cases, non-dowled bearings, etc., have had poor histories of performance.

j. Overhauled engine—in the event the engine has been overhauled, determine the extent of the overhaul. A top overhaul is only a partial engine refurbish process. Even a complete overhaul does not necessarily assure TBO. A remanufactured engine is in general the best of the engine rebuild processes.

2. Propeller:

a. Wood—check for evidence of delamination.

b. Metal—check for leading edge nicks. If greater than ¼ inch in diameter, the propeller may not be serviceable. Check constant-speed propellers for oil leakage at the hub.

3. Airframe:

a. Thorough airframe logbook review.

b. Thorough equipment list review.

c. Review of weight and balance data to assure currency.

d. Inspect all hydraulic shock struts, brake cylinders, landing gear actuators, and the like for evidence of leakage.

e. Inspect the airframe for evidence of damage or major repairs.

f. Test any fabric covering for service life. Select areas that have been constantly exposed to the sun or to water drainage (underside of fuselage) for test.

g. Test all avionics and instruments for valid operation by flying the aircraft.

During the flight test, be certain to check the aircraft for low static rpm or poor rate of climb. Both are indicative of a worn-out engine. If the aircraft is equipped with a constant-speed propeller, carefully cycle the prop through the full region of allowed manifold pressure and rpm as a test of both propeller and propeller governor operation. Any sign of power surging is serious and must be corrected before further flight. If the aircraft is a retractable, cycle the gear in the manual mode to assure proper emergency operation.

Having pushed all of the buttons and operated all of the levers, now you are in a position to make that buy/no-buy decision. At least you now know the condition of the aircraft and its accessories. It's now just a matter of price. But what is a going price? What is a fair value?

To answer the price question, begin by consulting your bank loan officer. A blue book exists for aircraft just as it does for cars. A simple phone call can provide you the high and low limits for your aircraft choice. Financing and interest rates? About all that can be said nowdays is shop around. Some banks simply will not make aircraft loans; others prefer not to. This is reflected in their interest rates or credit qualification requirements. Find the bank that has an aircraft loan officer or consider financing through an aircraft company such as Cessna. By dealing with a financial institution that understands the aircraft business, chances are best for the lowest interest rate or longest financing period. With the financial backing arranged, now buy and enjoy your purchase. You looked it over well, you know its good points and its deficiencies, you have a right to be proud of your careful selection. There should be no surprises. That flying vacation is about to become a reality at last.

Buying an Old Aircraft

Its tires were flat, the aluminum skin lay dulled with dirt and rain streaks, a faded sectional covered the instrument panel, the interior bathed in dust, the ailerons creaked with each puff of wind—but underneath there beat the heart of a pure thoroughbred—a Model 35 Bonanza. "A little clean-up, some soap and and water, a bit of polish and she'll be ready to go again. It's really a matter of elbow grease, and I've got plenty of that. At last I've found an aircraft which is a bargain, and they said it couldn't be done."

Now we must turn to another facet of the buying process, namely that there are indeed two basic types of pilots. Type A purchase an aircraft to fly, to use in his business, or to transport friends and family. Type B buys an aircraft to tinker with, to restore, to work on, for he is the craftsman of the fraternity. Type A flies in excess of 200 hours per year and farms out all of his maintenance. Type B flies about 25 hours per year and oversees every screw and nut his mechanic loosens or tightens in his aircraft. Where possible, he does all of his work to the limit allowed by FARs.

Buying an old aircraft? This author's advice to the Type A pilot is absolutely no. An aircraft that has been allowed to decay does so in *all* respects, airframe, engine, avionics, and instruments. Lack of maintenance or disuse is the first cousin to abuse. To the Type B pilot, the restoration of an antique can provide years of pleasure. Restore it to original or modernize it, whichever suits your fancy, for your pleasure is breathing life into an abandoned airframe, a neglected classic. It is to you that we owe a debt of gratitude for resurrecting from entombment those great airplanes of the past. To you, the old aircraft is indeed a bargain; to the Type A pilot, however, it is his millstone, his perpetual tormentor.

Other Bargain Pitfalls to Avoid

If you are a first-time aircraft buyer who is logically looking for an aircraft that is priced right, *beware of bargains*—certainly not a new warning for we realize this in almost all of the things we purchase daily. Then why mention it again? Why devote a paragraph to telling you what you already know? Just this: An aircraft isn't purchased every day. The elements of the "bargain" aircraft *are not* common knowledge, particularly to the first-time buyer. So let's spend a few moments looking at a few "bargains."

Lost service logs: Run from this one! Even if the price is unbelievably low, this one is a *definite no*! Don't even toy with the thought of buying an aircraft for which the service logs have been "lost."

Regrettably, it happened to an associate of mine. The engine quit on Jack over mountainous terrain with wife, child, and dog aboard. Plenty of altitude, some luck, and good piloting technique put the aircraft down without harm on a dirt road. A subsequent examination of the aircraft by the FAA disclosed the general engine condition to be a disaster from the maintenance standpoint. Furthermore, the engine had been removed from an aircraft listed as stolen. Net value of Jack's bargain—zero! 'Nuff said.

Off-brand models: If you are the type who enjoys owning an Edsel or considers the Corvair to be the car of the future, then ownership of an unusual or unique aircraft may be entirely satisfactory. If, however, you are a member of the Ford or GM crowd, be cautious. While unique aircraft may be perfectly serviceable, it's the resale or trade-in that becomes the problem. Examples of such are:

☐ Aircraft made by manufacturers that have ceased doing business. Parts and maintenance is a problem.

□ Aircraft with high-power engine conversions. Such aircraft are often converted for specialized flying reasons and may suffer disadvantages under routine service. Short range is a typical problem.

□ Aerobatic aircraft. Aerobatic aircraft are just that, and they are made for a special purpose. Don't expect this kind of machine to be a super cross-country aircraft.

□ Experimental aircraft (homebuilts). These machines are generally very well made but each is unique. Be prepared to do your own maintenance. Also remember that some of the experimentals have high landing speeds. Certain beautiful but short-winged models come over the fence at 100 and touch at 80 mph minimum. To be completely frank, many of the experimentals are simply not for the low-time pilot.

Some Lesser Bargain Problems

Looking toward other compromises that can provide room for either the buyer or seller to bargain, consider the following:

Seacoast aircraft: The majority of aircraft manufactured *do not* have corrosion protection from salt water (it's an option at the time of purchase). If the aircraft you are considering buying has been based and/or operated in seacoast areas, by all means have an experienced mechanic go over it for signs of corrosion.

Eighty octane engines: Like it or not, we are being forced to use 100 octane fuel to a greater extent every year. Certain of the older 80 octane aircraft engines do not take well to 100 octane fuel. Spark plug fouling and valve problems are common. While supplemental TCP may be added to reduce the effects of lead fouling, the task is somewhat of a nuisance. Further, the material must be handled with care. Older aircraft with 80 octane engines simply are worth less as 80 octane fuel becomes increasingly scarce.

Old avionics: The day of the vacuum tube aircraft radio is rapidly drawing to a close. In many cases it is most regrettable for much of our modern, semiconductor avionics equipment has failed miserably to provide the reliability which we as pilots need. The problem is simply that typical aircraft vacuum tube avionics equipment is approaching 20 years in age. End-life statistics are beginning to eat away at reliability. Further, some shops will not repair older equipment. Get an opinion from your local avionics repairman as to the value of the avionics in the aircraft you are considering and bargain accordingly.

Marginal IFR equipment: You wish to purchase an aircraft for training as a commercial pilot and eventual IFR flight. An excellent low-time aircraft is for sale but it appears light on avionics equipment. What is required as an absolute minimum?

The word "minimum" needs some interpretation as related to IFR flight. The *legal* minimum is one thing, but the *practical* minimum is another. Loss of your only nav-comm in a TCA during IFR conditions causes you to very quickly readjust your thinking as to minimum equipment needs for today's IFR environment. Frankly stated, IFR flight simply requires good equipment; there is no getting around it. I suggest the following as a *minimum* complement of IFR avionics:

☐ Primary nav-comm (720 channel).
☐ Secondary nav-comm (720 or 360 channel).
☐ Transponder (TSO'd).
☐ Encoding altimeter.
☐ Marker beacon receiver.
☐ Glide Slope receiver.
☐ Automatic Direction Finder.
☐ Headset and microphone with control column push-to-talk switch.
☐ Distance Measuring Equipment (optional).

In addition to minimum IFR avionics, good cockpit and instrument night lighting is a requirement. You simply *must* be able to read IFR charts and approach plates in a convenient fashion at night. A flashlight is a backup, *not* a primary means of night lighting.

AIRCRAFT OWNERSHIP

Now that you have made your aircraft selection, let's look at a few of the legalities of aircraft ownership to assure your purchase is without encumbrances. Let's also look at your responsibilities from the standpoint of maintenance and service logs, to protect your investment. The advice contained in the next few paragraphs has been abstracted from the FAA publication *Plane Sense*, AC 20-5D, and from the publication *FAA General Aviation News."*

Ownership Legalities

Q. *What is meant by a "clear title?"*

A. A "clear title" is one on which there are no encumbrances such as liens, chattel mortgages, or other claims against the aircraft.

Q. *How can I be sure that the aircraft has a "clear title?"*

A. Either search the aircraft records yourself, or have it done by an attorney or qualified aircraft title search company. You wouldn't think of purchasing a house until you had the title examined. You should do no less when purchasing an aircraft, which also represents a substantial investment. Even though you are planning to purchase the aircraft from an established dealer, it makes good sense to determine the true title status before you buy.

There is no substitute for examining the aircraft records to secure a history of the ownership of the aircraft and to determine if there are any outstanding liens or mortgages. This procedure will help avoid a delay in registering an aircraft and the headaches many have suffered because they failed to take this one important step before purchasing their aircraft.

Q. *Where do I go to search the records?*

A. All aircraft public records maintained by the U.S. Department of Transportation, Federal Aviation Administration (FAA), are on file at the FAA Aeronautical Center, Aviation Records Building, Aircraft Registration Branch, AAC-250, 6500 South MacArthur Boulevard, Oklahoma City, Oklahoma 73125.

Q. *What documents may I expect to receive with my new or used aircraft?*

A. 1. Bill of Sale.

   2. Either a Standard Airworthiness Certificate, FAA Form 8100-2; or a Special Airworthiness Certificate, FAA Form 8130-7.

   3. Maintenance records containing the following information:

     a. The total time in service of the airframe;

     b. The current status of life-limited parts of each airframe, engine, propeller, rotor, and appliance;

     c. The time since last overhaul of all items installed on the aircraft that are required to be overhauled on a specified time basis;

     d. The identification of the current inspection status of the aircraft, including the times since the last inspections required by the inspection program under which the aircraft and its appliances are maintained;

     e. The current status of applicable airworthiness directives, including the method of compliance;

     f. A list of current major alterations to each airframe, engine, propeller, rotor, and appliance.

4. Equipment lists, weight and balance data.

5. Maintenance manuals, service letters, bulletins, etc.

6. Airplane Flight Manual or operating limitations.

Q. *Does a current 100-hour or annual inspection mean that the aircraft is in "first-class" condition?*

A. No. It indicates only that the aircraft was found to be in a condition for safe operation at the time of inspection.

Q. *What should I do before buying an amateur-built or experimental aircraft?*

A. Contact the General Aviation or Flight Standards District Office serving your locale and ask to speak to an airworthiness inspector who will explain the requirements for experimental certification.

Q. *What should I consider when buying a military surplus aircraft?*

A. Certain military surplus aircraft are not eligible for FAA certification in the Standard, Restricted, or Limited classifications. Since no civil aircraft may be flown unless certificated, you should discuss this with the local FAA inspector, who will advise you of eligible aircraft and certification procedures.

Q. *What are my responsibilities as an aircraft owner?*

A. As an aircraft owner, you will be assuming responsibilities similar to those you have if you own an automobile. Owning an automobile usually means that you must register it in your state of residence and obtain license plates. As the registered owner of an aircraft, you will be responsible for:

1. Having a current Airworthiness Certificate and Aircraft Registration Certificate in your aircraft.

2. Maintaining your aircraft in an airworthy condition.

3. Assuring that maintenance is properly recorded.

4. Keeping abreast of current regulations concerning the operation and maintenance of your aircraft.

5. Notifying the FAA Aircraft Registry immediately of any change of permanent mailing address, of the sale or export of your aircraft, or of the loss of your U.S. citizenship.

Some states require that your car be inspected periodically (typically, every six to twelve calendar months) to assure that it is in a safe operating condition. Your aircraft will have to be inspected in accordance with an annual inspection or with one of the five inspection programs outlined in Federal Aviation Regulations Part 91,

Section 91.217, in order to maintain a current Airworthiness Certificate. As with your automobile, accidents involving your aircraft must be reported.

Some similarities between automobile and aircraft responsibilities are shown in the following chart:

Responsibility	Automobile	Aircraft
Registration	Yes	Yes
Inspection	Yes	Yes
Compulsory Insurance (most states)	Yes	No
Reporting of accidents	Yes	Yes
Required Maintenance records	No	Yes
Maximum speed restrictions	Yes	Yes
Controlled maintenance	No	Yes

Q. *How do I obtain a Certificate of Registration and Bill of Sale?*

A. If you purchase an aircraft, you must apply for a Certificate of Aircraft Registration from the FAA Aircraft Registry before you can legally fly it. An aircraft is eligible for registration only if it is owned by a citizen of the United States or a governmental unit and is not registered under the laws of any foreign country.

You may obtain an Aircraft Registration Application , AC Form 8050-1, consisting of an original (white) and two duplicate copies (green and pink) from an FAA General Aviation District Office. Instructions for preparing and submitting the form are attached to it.

When applying for a Certificate of Aircraft Registration, you must also submit an aircraft bill of sale or other evidence of ownership. A bill of sale that meets the recording requirements of the Federal Aviation Administration is AC Form 8050-2, Aircraft Bill of Sale.

Until you receive the permanent Certificate of Aircraft Registration, AC Form 8050-3, from the FAA, the pink copy of the application serves as a temporary certificate for 90 days and must be carried in the aircraft. The Certificate of Aircraft Registration replaces the pink copy of the application in the aircraft.

Aircraft previously registered in a foreign country may also be operated using the pink copy of the application as a temporary registration certificate, if the applicant carefully follows the steps outlined on page 80.

1. Request that a registration number be assigned by submitting the make, model, and serial number of the aircraft, with the confirmation of foreign registration cancellation, to the FAA Aircraft Registry.
2. If the foreign registration has not ended, submit an affidavit stating that the U.S. registration number will not be placed on the aircraft until the foreign registration has been cancelled.
3. After a U.S. registration number has been assigned by the Registry, you have 90 days to complete the requirements for registration by forwarding evidence of ownership, white and green copies of the application, registration fee, and confirmation of cancellation from the foreign registry.
4. As soon as the foreign registration has been cancelled, you may place the U.S. registration number on the aircraft, mail the above documents, and use the pink copy of the application, as with any other U.S. civil aircraft.

The Certificate of Aircraft Registration will expire when:

1. The aircraft is registered under the laws of a foreign country;
2. The registration of the aircraft is cancelled at the written request of the owner;
3. The aircraft is destroyed or scrapped;
4. The ownership of the aircraft is transferred;
5. The holder of the certificate loses his United States citizenship; or
6. Thirty days have elapsed since the death of the holder of the certificate.

When the aircraft is destroyed, scrapped, or sold, the previous owner must notify the FAA by filling in the back of the Certificate of Aircraft Registration and mailing it to the FAA Aircraft Registry.

When a U. S. civil aircraft is transferred to a person who is not a U. S. citizen, the U. S.-registered owner is required to remove the United States registration and nationality marks from the aircraft before the aircraft is delivered.

A Dealer's Aircraft Registration Certificate is another form of registration certificate, but it is valid only for required flight tests by the manufacturer or in flights that are necessary for the sale of the aircraft by the manufacturer or a dealer. It must be removed by the dealer when the aircraft is sold.

The FAA does not issue any certificate of ownership or endorse any information with respect to ownership on a Certificate of Aircraft Registration.

Q. *What is an Airworthiness Certificate?*

A. An Airworthiness Certificate is issued by a representative of the Federal Aviation Administration after the aircraft has been inspected, is found to meet the requirements of the Federal Aviation Regulations (FAR), and is in a condition for safe operation. The Certificate must be displayed in the aircraft so that it is legible to passengers or crew whenever the aircraft is operated. The Airworthiness Certificate is transferred with the aircraft, except when it is sold to a foreign purchaser.

The Standard Airworthiness Certificate, FAA Form 8100-2, is issued for aircraft-type certificated in the normal, utility, acrobatic, and transport categories, or for manned free balloons.

A Standard Airworthiness Certification remains in effect as long as the aircraft receives the required maintenance and is properly registered in the United States.

If you are interested in purchasing an aircraft classed as other than Standard, it is suggested that you contact the local FAA General Aviation or Flight Standards District Office for an explanation of the pertinent airworthiness and the limitations of such a certificate.

In summary, the FAA initially determines that your aircraft is in a condition for safe operation and conforms to type design, then issues an Airworthiness Certificate.

AIRCRAFT MAINTENANCE, OWNER RESPONSIBILITIES

Maintenance means the inspection, overhaul, and repair of aircraft, including the replacement of parts. *A properly maintained aircraft is a safe aircraft.* The purpose of maintenance is to ensure that the aircraft meets acceptable standards of airworthiness throughout its operational life.

Although maintenance requirements will vary for different types of aircraft, experience shows that most aircraft will need some type of preventive maintenance every 25 hours of flying time or less, and minor maintenance at least every 100 hours. This is influenced by the kinds of operation, climatic conditions, storage facilities, age, and construction of the aircraft. Most manufacturers supply service information which should be used in maintaining your aircraft.

Inspections

FAR Part 91 places primary responsibility on the owner or operator for maintaining an aircraft in an airworthy condition. Certain inspections must be performed on your aircraft, and you must maintain the airworthiness of the aircraft between required inspections by having any defects corrected.

Federal Aviation Regulations require the inspection of all civil aircraft at specific intervals to determine the overall condition. The interval depends generally upon the type of operations engaged in. Some aircraft need to be inspected at least once each 12 calendar months, while inspection is required for others after each 100 hours of operation. In other instances, an aircraft may be inspected in accordance with an inspection system set up to provide for total inspection of the aircraft on the basis of calendar time, time in service, number of system operations, or any combination of these.

Annual Inspection: Any reciprocating-powered light aircraft, 12,500 pounds and under, flown for pleasure is required to be inspected at least annually by a certificated airframe and powerplant mechanic holding an inspection authorization, or a certificated repair station that is appropriately rated, or the manufacturer of the aircraft. The aircraft may not be operated unless the annual inspection has been performed within the preceding 12 calendar months. A period of 12 calendar months extends from any day of any month to the last day of the same month the following year. However, an aircraft with the annual inspection overdue may be operated under a special flight permit for the purpose of flying the aircraft to a location where the annual inspection can be performed.

100-Hour inspection: Any reciprocating-power light aircraft, 12,500 pounds and under, used to carry passengers or for flight instruction for hire must be inspected within each 100 hours of time in service by a certificated airframe and powerplant mechanic, a certificated repair station that is appropriately rated, or the aircraft manufacturer. An annual inspection is acceptable as a 100-hour inspection, but the reverse is not true.

Other inspection programs: The annual and 100-hour inspection requirements do not apply to large airplanes, turbojet, or turboprop-powered multiengine airplanes, or to airplanes for which the owner or operator complies with the progressive inspection requirements. Details of these requirements may be determined by reference to Parts 43 and 91 of the Federal Aviation Regulations and by inquiry at a local FAA General Aviation or Flight Standards District Office.

Preventive Maintenance

Simple or minor preservation operations and the replacement of small standard parts, not involving complex assembly operations, are considered *preventive maintenance*. Certificated pilots may perform preventive maintenance on any aircraft owned or operated by them that are not used in air carrier service. FAR Part 43 lists the following as preventative maintenance work which the pilot-owner may accomplish:

☐ Removal, installation, and repair of landing gear tires.

☐ Replacing elastic shock absorber cords on landing gear.

☐ Servicing landing gear shock struts by adding oil, air, or both.

☐ Servicing landing gear wheel bearings, such as cleaning and greasing.

☐ Replacing defective safety wiring or cotter keys.

☐ Lubrication not requiring disassembly other than removal of nonstructural items, such as cover plates, cowlings, and fairings.

☐ Making simple fabric patches not requiring rib stitching or the removal of structural parts or control surfaces.

☐ Replenishing hydraulic fluid in the hydraulic reservoir.

☐ Refinishing decorative coating of fuselage, wings, tail group surfaces (excluding balanced control surfaces), fairings, cowling, landing gear, cabin, or cockpit interior when removal or disassembly of any primary structure or operating system is not required.

☐ Applying preservative or protective material to components where no disassembly of any primary structure or operating system is involved and where such coating is not prohibited or is not contrary to good practices.

☐ Repairing upholstery and decorative furnishings of the cabin or cockpit interior when the repairing does not require disassembly of any primary structure or operating system or interfere with an operating system or affect primary structure of the aircraft.

☐ Making small simple repairs to fairings, nonstructural cover plates, cowlings, and small patches and reinforcements not changing the contour so as to interfere with proper airflow.

☐ Replacing side windows where that work does not interfere with the structure or any operating system such as controls, electrical equipment, etc.

☐ Replacing safety belts.

☐ Replacing seats or seat parts with replacement parts approved for the aircraft, not involving disassembly of any primary structure or operating system.

☐ Troubleshooting and repairing broken circuits in landing light wiring circuits.

☐ Replacing bulbs, reflectors, and lenses of position and landing lights.

☐ Replacing wheels and skis where no weight and balance computation is involved.

☐ Replacing any cowling not requiring removal of the propeller or disconnection of flight controls.

☐ Replacing or cleaning spark plugs and setting of spark plug gap clearance.

☐ Replacing any hose connection except hydraulic connections.

☐ Replacing prefabricated fuel lines.

☐ Cleaning fuel and oil strainers.

☐ Replacing batteries and checking fluid level and specific gravity.

☐ Removing and installing glider wings and tail surfaces that are specifically designed for quick removal and installation and when such removal and installation can be accomplished by the pilot.

FAR 43.13 requires that preventive maintenance must be done using methods, techniques, tools, equipment, and test apparatus necessary to assure completion of the work in accordance with accepted industry practices. If special equipment or test apparatus is recommended by the manufacturer involved, then that equipment or apparatus or its equivalent must be used. The work performed must be done in such a manner and the materials used must be of such quality that the condition of the aircraft, airframe, aircraft engine, propeller, or appliance worked on will be at least equal to its original or properly altered condition. Pilots should exercise caution before engaging in preventive maintenance, since many of today's aircraft are highly sophisticated and complex machines. What may appear to be a simple task may turn out to be a highly complicated operation. Therefore, carefully consider if the intended operation is within your capability before you begin the work.

Repairs and Alterations

Except as noted under the previous paragraph, Preventive Maintenance, all repairs and alterations are classed as either major or minor. *Major* repairs or alterations must be approved for return to service by an appropriately rated certificated repair station, an airframe and powerplant mechanic holding an inspection authorization, or a representative of the FAA. *Minor* repairs and alterations may be returned to service by a certificated airframe and powerplant mechanic or an appropriately certificated repair station.

Airworthiness Directives

A primary safety function of the Federal Aviation Administration is to require correction of unsafe conditions found in an aircraft, aircraft engine, propeller, or appliance when such conditions exist and are likely to exist or develop in other products of the same design. The unsafe conditions may exist because of a design defect, maintenance, or other causes. FAR Part 39, Airworthiness Directives, defines the authority and responsibility of the Administrator for requiring the necessary corrective action. The Airworthiness Directives (ADs) are the medium used to notify aircraft owners and other interested persons of unsafe conditions and to prescribe the conditions under which the product may continue to be operated (Fig. 5-1).

Airworthiness Directives may be divided into two categories: those of an emergency nature requiring *immediate* compliance upon receipt, and those of a less urgent nature requiring compliance within a relatively longer period of time.

Airworthiness Directives are Federal Aviation Regulations and must be complied with, unless specific exemption is granted. It is the aircraft owner's or operator's responsibility to assure compliance with all pertinent ADs. This includes those ADs that require recurrent or continuing action. For example, an AD may require a repetitive inspection each 50 hours of operation, meaning the particular inspection shall be accomplished and recorded *every* 50 hours of time in service.

Federal Aviation Regulations require a record to be maintained that shows the current status of applicable airworthiness directives, including the method of compliance, and the signature and certificate number of the repair station or mechanic who performed the work. For ready reference, many aircraft owners have a chronological listing of the pertinent ADs in the back of their aircraft and engine records.

The Airworthiness Directives Summary contains all the valid ADs previously published and biweekly supplements. The Summary is divided into two volumes. Volume I includes directives applicable to small aircraft (12,500 pounds or less maximum certificated takeoff weight). Volume II includes directives applicable to large aircraft (over 12,500 pounds). Subscription service consists of

EMERGENCY AIRWORTHINESS DIRECTIVE
DEPARTMENT OF TRANSPORTATION
FEDERAL AVIATION ADMINISTRATION

FLIGHT STANDARDS SERVICE
FLIGHT STANDARDS NATIONAL FIELD OFFICE
P.O. BOX 25082
OKLAHOMA CITY OKLAHOMA 73125

February 4, 1980

Pursuant to the authority of the Federal Aviation Act of 1958, delegated to me by the Administrator, the following Airworthiness Directive (AD) is issued and applicable to all owners and operators of aircraft with magnetos manufactured by Slick Electro, Inc., Rockford, Illinois with model and serial numbers as follows:

MODEL NO.	RANGE OF APPLICABLE MAGNETO SERIAL NUMBERS*
447	9040001 thru 9040049
447R	"
662	9020462 thru 9070000
662R	"
664	9040001 thru 9040086
664R	"
680	9020462 thru 9070000
680R	"
4151	9020017 thru 9070000
4151R	"
4152	"
4152R	"
4181	"
4181R	"
4201	9020210 thru 9070000
4201R	"
4230	9040001 thru 9040197
4230R	"
4251	9030001 thru 9070000
4251R	"
4281	"
4281R	"
6210	8090073 thru 9070000
6214	8050001 thru 9070000

*Any magneto serial number between and including the lower and upper numbers shown are affected by this AD.
These magnetos are installed on, but not limited to, the following engines:

LYCOMING	AEIO-360
	AEIO-320
	IO-320
	O-235
	O-320
	O-360

EMERGENCY AIRWORTHINESS DIRECTIVE

Fig. 5-1. The Airworthiness Directive is the manner in which the FAA advises aircraft owners of equipment defects and prescribes corrective action.

the summary and automatic biweekly updates to each summary for a two-year period. The Summary of Airworthiness Directives, Volume I and Volume II, are sold and distributed for the Superintendent of Documents by the FAA from Oklahoma City. Requests for subscription prices to either of these publications should be sent to:

U. S. Department of Transporation
Federal Aviation Administration
Aeronautical Center
Attention: AAC-23
P. O. Box 25461
Oklahoma City, Oklahoma 73125

As an aircraft owner, pilot, or certificated mechanic, you may at some time want to obtain copies of documents pertaining to your aircraft, airman, or medical certificates. Copies of aircraft and airman records are 5¢ for each page. A $2.00 fee is charged for searching records and furnishing duplicate original documents. Duplicate airman certificates, medical certificates, or certificates of aircraft registration are $3.00. Fees, which are subject to change, may be paid by check, draft, or postal money order, made payable to the Federal Aviation Administration and submitted to the Aircraft Registration Branch, AAC-250; Airman Certification Branch, AAC-260; or Aeromedical Certification Branch, AAC-130, as appropriate, at the following address:

U. S. Department of Transportation
Federal Aviation Administration
FAA Aeronautical Center
P. O. Box 25082
Oklahoma City, Oklahoma 73125

Service Logs

The paperwork involved in keeping up a good and thorough set of aircraft records can be an unappealing chore, but it is one which can ultimately save you time, money, and perhaps even your life. Furthermore, it is a task which you cannot always delegate to others. The businessman who is used to having a secretary to write his letters, an accountant to do his bookkeeping, and a mechanic to maintain his aircraft, may assume that the latter will also be responsible for maintenance record keeping. But this is sometimes a false assumption, which could cost him.

Consider what happened on these occasions when a pilot/ owner failed to manage his aircraft records properly:

An Airworthiness Directive was issued for a Mooney 21 requiring the replacement of the connecting rods and assembly pins. It was an expensive repair (since the engine had to be extensively disassembled to get at the connecting rods) but it had to be done. The owner paid his bill—something over $400—and flew off in his airplane, eager to be gone. It happened that he had relocated to a different part of the country when the time for his next annual inspection came due, almost a year later. The mechanic who was to perform the inspection reviewed the list of applicable ADs for that particular airplane, but could find no record of compliance with the AD concerning connecting rods. The pilot insisted the work had been done; he even had his cancelled check, but the original mechanic could not be located, *and there was no written record of the compliance*. The airplane was not legally airworthy. The engine had to be disassembled again to make sure the connecting rods and assembly pins had been inserted. And the chagrined pilot had to foot the bill twice for virtually the same job—all because he did not take time to confirm that he had a written description of the work as performed, signed by the mechanic. (The latter was also at fault for failing to record the work.)

The FARs, essentially sections of Parts 91 and 43, contain specifics for the kind of aircraft records which must be kept, and how and by whom they must be maintained. In many accidents where these regulations about record keeping have not been followed, the insurance company has been successfully able to avoid any renumeration!

Q. *Who is responsible for the aircraft records?*

A. Actually, the responsibility is shared. Whenever a mechanic performs maintenance or alterations on an aircraft, he is required to furnish the owner or operator with a brief written description of what was done. His signature, together with his FAA certificate number, indicates that the aircraft is approved for return to service (FAR 43.9). But the *owner* or *operator* is responsible (FAR 91.173) for making certain that a proper record of such service is maintained. The phrase "owner or operator" here is understood to mean the party contractually responsible for maintaining the aircraft at a given time—where an aircraft is leased, for example, the person leasing it could be the *operator* and have the responsibility, rather than the owner. Therefore, even though a mechanic may be in violation for failing to make a proper entry, the owner or operator of the aircraft can also be in violation if the aircraft records are not in order.

On some occasions the pilot may be neither the owner nor the operator, but he nevertheless has responsibility for the airworthiness of the aircraft. Part 91.29 of the regulation states that *"No person may operate a civil aircraft unless it is in an airworthy condition. The pilot-in-command . . . is responsible for determining whether that aircraft is in condition for safe flight.*" This does not mean that he must read the service record before each flight, but is does mean that he should know the status of inspections and recent maintenance.

Furthermore, FAR 43.5 speaks directly to the owner, operator and pilot when it states that *"No person may return to service any aircraft, airframe, aircraft engine, propeller or appliance that has undergone maintenance, preventive maintenance, rebuilding, or alteration unless it has been approved for return to service by a person authorized under Part 43.7, and the required maintenance record entry has been made."* This means that even if the airplane only went into the shop for an oil change, it may not be legally flown until the mechanic has approved it for return to service *in writing*.

Q. *What records do you have to keep?*

A. There are two types of records which the regulations oblige you to keep for the three major components of the aircraft—the airframe, each engine, and each propeller or rotor.

I. Temporary Records
   a. A record of all minor maintenance and minor alterations performed on the aircraft by mechanics.
   b. A record of the required inspections performed on the aircraft, whether it be the 100 hour, or the annual, or a progressive inspection, or any other required or approved inspection.

These records are required by FAR 91.173, revised in 1972. Minor maintenance and alterations records may be discarded when the work has been repeated, or superceded by other work, or in any case, one year after the work has been performed. The record of routine inspections may also be discarded when the next inspection is completed. (Many owners choose to retain *all* their aircraft records indefinitely, for the sake of a more complete record, which is good practice.)

II. Permanent Records
   a. Total time in service of the airframe.
   b. The current status of the life-limited parts of each airframe, engine, propeller, rotor, and appliance.
   c. The time since the last overhaul of items on the aircraft which are required to be overhauled on a scheduled time basis.
   d. The current inspection status of the aircraft.
   e. The current status of applicable Airworthiness Directives and the methods of compliance.

f. A list of the current major alterations to each airframe, engine, propeller, and rotor.

g. Current operating limitations, including revisions to the aircraft weight and center of gravity, caused by the installation or removal of equipment or alterations. (FAR 91.31 and 43.5).

Q. *What kind of format is required?*

A. The regulations do not specify any particular format for your aircraft records. You may develop your own system, as long as it includes all the necessary data. You may use a separately bound log, for example, for the airframe, another for the engines, a third for propeller or rotors. Or you may keep your service records consecutively in a loose-leaf book. Obviously, the more clearly and systematically this information is recorded, the easier it will be to consult.

Q. *How are entries made in those records?*

A. Part 43 describes two main types of entries in aircraft records: those for regular maintenance, alterations and repairs, and those for inspections.

a. The first type of entry is for any work by a mechanic which involves maintaining or rebuilding or altering either an aircraft, an airframe, an engine, a propeller, or appliance. The entry must include these items: (FAR 43.9)

☐ A description of the work (or reference to some data acceptable to FAA).

☐ The date the work was completed.

☐ The mechanic's name.

☐ If approved for return to service, the signature and certificate number of the mechanic approving it.

b. The second type of entry is made when a mechanic approves or disapproves an aircraft for return of service after an annual, 100 hour, or progressive inspection. This entry must include: (FAR 43.11)

☐ The type of inspection.

☐ The date of the inspection.

☐ The aircraft time in service.

☐ The signature and certificate number of the mechanic approving or disapproving for return to service.

Q. *Where do I keep the records?*

A. The regulations do not dictate where the aircraft maintenance records should be kept, just that they must be available to

FAA or NTSB if requested. Practicality, however, dictates that they should be conveniently located so that the mechanic can review them before he does any work on the airplane, and he can sign them promptly after he finishes. For this reason, many private owners do keep their aircraft logbooks in the airplane for ready availability, especially if repairs have to be made out of town. Other owners prefer to keep their records at home in a file. Maintenance records for rental aircraft are frequently kept in the FBO's office.

AIRCRAFT OWNERSHIP COSTS

The final consideration is cost. Whether for business, training, or pleasure, what will the cost be? Until this time, the pleasure of seeking an aircraft, weighing equipment tradeoffs, and planning a series of long-awaited trips has been pleasure extraordinary. Buying fever has been a delightful malady. But now comes shock treatment, the cost of ownership! "Is it really going to be worth it? Can I justify the expenditure? What if I'm transferred to another area? Business is okay now, but what about next year? Maybe the family won't like flying. What if . . ." and on and on the list of negative suppositions grows. Better think this thing over, better take more time. "But the plane might sell to someone else," torments that inner voice. What is the correct thing to do? How can you be assured that you won't regret your decision latter? The final step in the decision process, *the bottom line*, is ownership cost.

It's time for a bit of very frank talk on the subject of ownership costs. To put the matter in the proper perspective, consider carefully the following three facts:

1. You get exactly *one* go-around in this life—no more!
2. The thrill of flying and aircraft ownership is not something that can be delayed until old age to start to enjoy. Flying requires a healthy body and mind.
3. Flying is expensive—let's be absolutely honest about that fact. If you truly enjoy flying, something else is going to have to give (unless you are independently wealthy or the aircraft is a pure business investment).

Now that the "what ifs" have been put aside, let us look directly at the costs of aircraft ownership and begin a flying budget accordingly. The following lists typical cost factors for 1981; adjust accordingly for inflation if need be.

Fixed Costs

The fixed cost of flying are those that go on whether the aircraft is flown or not. For most pilots of single engine aircraft, these will be:

Storage
    Tiedown: Typically $10 to $25 per month.
    Hangar: Typically $70 to $175 per month;

Insurance*:
    Liability: About the same as a car.
    Hull: Typically 2% to 6% of aircraft value.

Taxes
    State: Consult your local State Department of Aeronautics.
    Federal: Currently $25 plus 2¢ per pound over 2500 lbs.

Required Maintenance
    Annual: Typically $200 to $500
    Transponder: Typically $25 to $50   } IFR only
    Pitot-Static: Typically $100 to $200
Aircraft Payment: Per contract

Variable Costs

The variable costs of flying are generally considered to be those that are accrued on a per-flying-hour basis. Again, referring to a typical single engine aircraft, these may be categorized as follows:

Gasoline:	Typically	$ 1.75 per gallon.
Oil:	Typically	$ 1.50 per quart.

*Rates are variable depending upon aircraft age, pilot experience, ratings, no. of pilots, etc. For accurate information, contact a major insurance company such as Avemco, of Bethseda, Maryland, or Don Flower Associates of Wichita, Kansas.

Maint. Kitty*:

Engine:	Suggest $ 5.00 per hr. min.
Avionics:	Suggest $ 2.00 per hr. min.
Instruments:	Suggest $ 1.00 per hr. min.
Airframe:	Suggest $ 1.00 per hr. min.
Prop and Miscellaneous	Suggest $ 1.00 per hr. min.
Maintenance Total	= $10.00 per hr. min.

Other Flying Costs

The following three items list costs which are simply a part of flying:

Physical Exam:	Typically $30.
Biennial Flt. Review:	Typically $20 to $50.
Flight Training:	
Flight Instructor:	Typically $15 to $25 per hour.
Rented C 152:	Typically $25 to $30 per hour.
Rented C 182 RG:	Typically $50 to $65 per hour.
Rented C 310:	Typically $120 to $150 per hour.

As we view the cost of aircraft rental, the value of owning one's own aircraft begins to make more sense. If we are going to fly, we may as well build up an equity in our own machine rather than pay for someone else's. Now, add up those costs for your selected aircraft and see how the budget looks. Unless you know how many hours per year you will be flying, use the national average of 75 hours per year as a base for your estimate. Now that a cost base has been established, let's see what we can do to reduce expenses.

SPREADING THE COST AROUND

The fact that flying is expensive should not be allowed to deny us our aviation goals. Indeed, if a person wants badly enough, he will find a way to achieve his aims despite the obstacles. "Rubbish," you say, "Dollars are dollars—you can't spend what you don't have. How can I expect to own a $40,000 airplane on a salary of $25,000 per year—with a wife and family to support? No, aircraft ownership is not for me; it's for the other guy. That other guy, just what does he have that I don't?"

The answer to this question is very straightforward. That other guy *put his ingenuity to work*. He looked for and found aviation

---

*The object of a maintenance kitty is to put aside money on a regular basis for engine overhaul, avionics repair, etc. Costs are typical for a mid-time engine and a recent model aircraft.

business associations or other means for cost sharing. Pilots, being basically ingenious creatures, aren't to be stopped by mere dollars, so let's have a look at what can be done to get in the air at a reasonable cost.

Business Association

Without doubt, one of the best means of minimizing out-of-pocket expenses for aircraft ownership and operation is to have a bona fide business reason for flying. This places the aircraft in the position of being an item of capital equipment, just as a building, a mill, or some other piece of machinery that is necessary to conduct the business. Depending upon the exact nature of the business, the following may qualify for tax benefits:

- ☐ Investment credit.
- ☐ Depreciation.
- ☐ Taxes paid on aircraft.
- ☐ Insurance premiums.
- ☐ Interest on payments.
- ☐ Tiedown or hangar rental.
- ☐ Maintenance expenses.
- ☐ Fuel and oil used.
- ☐ Travel expenses.
- ☐ Flight training expenses.

To pursue this avenue of financing an aircraft, consult a Certified Public Accountant (CPA) *who is knowledgable in aircraft accounting*. The field of flying is specialized and the average tax accountant has little contact with the art. An aircraft-oriented CPA can advise as to what constitutes a bona fide flying business association as well as the applicable tax benefits. After all, a reduction in taxes is just as beneficial as a comparable increase in income.

Aircraft Leaseback

The purchase of an aircraft and leaseback to the seller for charter or flying school use is one means of establishing a business arrangement for your aircraft. As business arrangements go, the leaseback is one of the higher-risk businesses to enter into. It's easy to start but difficult to terminate, without taking a loss, should you decide that you wish to stop leasing your aircraft. Before entering into a leaseback agreement, consider the following guidelines to doing business:

    1. The only real profit you will earn will be in the form of tax

benefits (see previous paragraph). Therefore, you *must* be in a tax bracket that will make these benefits worthwhile to you.

2. Recognize that the leaseback earns best only during the period the aircraft is covered by the manufacturer's warranty on avionics, engine, and airframe parts, and maintenance. Your earnings may well drop to zero (or less) after the warranty period expires. As the fixed base operator will generally insist that he determine what maintenance the aircraft will have, this can leave you in a vulnerable position after that first year of ownership.

3. A good fixed base operator may well put 80 to 100 hours per month on your aircraft. This means you will have at least a mid-time or possibly a high-time engine in only a year or so. What then? The aircraft is reaching a low value, comparatively, as a salable item, particularly if the engine is approaching 1400 hours or so. Furthermore, many fixed-base operators will not leaseback an aircraft for more than 18 to 24 months. They must have new aircraft on the line to do business.

One option is to trade the aircraft in every year and start the leaseback process over again. When the day comes that you wish to terminate the leaseback business, expect to take a loss. The loss can be minimized by your using the leaseback aircraft as a final trade on the aircraft you eventually wish to buy for your personal use.

4. Leaseback *only* the following two types of aircraft:

    a. Trainer aircraft, such as a Cessna 152 or equivalent. Being a trainer, the oil level, etc., will get checked routinely. The fact that trainer aircraft have a short range and are used fundamentally for local flights is a definite advantage in reducing your investment risk. Don't worry about student damage to the aircraft—it's really a rare event in practice.

    b. Aircraft that are to be flown *only* by company pilots. In the case of larger aircraft, there is only one way to protect your business investment—contract that your aircraft be flown *only* by authorized company pilots. Regrettably, larger aircraft that can carry a good payload fast and over long ranges have strange ways of showing up in Mexico, Columbia, and other obscure places for dubious reasons if simply rented to the general public—a deplorable act, a pitiful use for an aircraft, but a true fact of leasing, nevertheless.

One means of controlling an aircraft to be rented is to establish your own leasing company. The practice is really quite common.

Consult your aviation-oriented CPA as to requirements for establishing a sole proprietor, partnership, or possibly a corporate business structure in your state. Insurance companies are set up to handle this type of flying activity and can often offer good advice as to how to qualify your selected renter pilots from the standpoint of minimizing insurance premiums.

Partnerships

One means of splitting the fixed costs of aircraft ownership is to establish a partnership. This technique works out well up to a maximum of four partners. Beyond this point, the activity becomes a club.

In establishing a partnership, *absolutely* have a lawyer draw up a legal partnership agreement. You have elected to own property in companionship with another and are responsible for the actions of the other party. A properly constructed partnership agreement will define your liability for the actions of your partner. If your partnership numbers three or four persons, consider incorporation as a means of limiting your individual liability. Whatever technique you choose, get started on the right foot by establishing a carefully considered legal agreement as to the terms of the joint ownership.

Can partnership work well in practice? Some work very well, some create much bitterness. Experience shows that poor partnerships can be avoided by simply considering a few basic circumstances before signing the partnership agreement. Typically, some of the more obvious conditions are as follows:

1. All partners *must* be financially capable and *must* pay their share on time. Payment in advance (for fixed costs) is a good practice to establish.

2. All partners must agree *absolutely* on a priority for sharing the use of the aircraft.

3. All partners must agree *absolutely* on the maintenance program for the aircraft. Who will clean and polish the aircraft? Who will accomplish the maintenance tasks; what repair facility will be used? How will engine overhaul time be determined? Who will maintain the aircraft records? (This includes checking *all* partners to ensure that they are current in their physical exams and biennial flight reviews.)

4. All partners must agree on the extent of insurance coverage and their individual liability if they damage the aircraft.

5. All partners must agree on the terms of a partner selling his share and the acceptance of a new partner by the remainder of

the partnership. This same type of consideration necessarily applies in the event of a death or incapacitation of a partner.

6. All partners must agree *absolutely* on who shall fly the aircraft. Example: A, B, and C each agree to a ⅓ interest in a Cessna 182. However, in family C the father, wife, and two sons are all licensed pilots. All expect to fly on C's ⅓ interest. Say no to this one! Partnerships are between *individuals*, not families.

7. Partners must agree on upgrading equipment as a part of the partnership agreement. Example: A, B and C own equal shares of an older model Cessna 210. A and B regularly fly IFR and agree that the radio must be replaced in order to provide a reliable ILS capability. C is strictly a VFR pilot and has no use for an increased instrument capability. Who pays for the equipment upgrade?

8. The partnership agreement must contain terms for dissolving the agreement as well as conditions for dismissing a partner. Example: A, B, and C has a splendid partnership until B's wife found a new playmate. B must be terminated from the partnership in order that the aircraft not become a part of B's divorce and property settlement.

Perhaps the most significant feature of a partnership is that each partner realize that he or she must give up a little bit of independence in the use of an aircraft in order to gain a lot of financial benefit. Compatibility of the persons involved is always the key to a successful partnership. It can be done. After all, there are many who are seeking a way to fly at a reduced cost.

Flying Clubs

Flying clubs come in two varieties, the private flying club and the aviation operator-sponsored flying club. Without question, the flying club is one of the least expensive ways to get a start in the flying activity and share in a piece of an aircraft, however small.

A well-organized private flying club is incorporated, has a regular flight instructor on call, has been in existence several years, and has at least two aircraft available. Typically, private flying clubs number ten to twelve persons per aircraft, have a telephone scheduling service, and established rules and regulations pertaining to membership and aircraft usage. An initial membership fee of $300 plus monthly dues of $25 to $50 is not unusual.

As strange as it may seem, one of the common problems of a flying club is not charging enough. The scenario goes something like this: An aircraft is purchased by a group and a club subsequently formed. The aircraft is operated with just enough income to cover day-to-day expenses. A major repair is needed and there has been inadequate money set aside. In effect, the club is bankrupt, the aircraft is grounded, and all flying activity ceases. The aircraft must now be sold at a substantial loss and the net shared between the members. All lose significantly on their investment.

Let's face the facts once again: there is no such thing as flying for $5.00 per hour (unless the club owns only an ultralight and perhaps one hang glider). Find a club that has a *proven* record of financial solvency and enjoy your membership therein. Don't expect the aircraft to be pristine in terms of cleanliness and polish. Private clubs simply don't work that way.

In looking for a private flying club, be absolutely certain to determine whether or not the club is incorporated. If it isn't, you have just become a party to a joint ownership in an aircraft with a large number of other joint owners. Say no to this proposition; it isn't worth the risk.

Flying clubs of a commercial nature are often sponsored by a fixed base operator or aircraft manufacturer. These clubs offer a reduction in flying rates in return for your continued business. Typically, it is necessary to fly about four hours per month in order to break even, that is, offset the monthly dues against the reduced aircraft rental rates offered you.

One of the finest of the commercial organizations is the Beech Aero Club. Beech has the following to say about their club:

"Maybe you're not ready for aircraft ownership right now. But that doesn't mean you can't have access to one. Membership in the Beech Aero Club puts a fleet of fine Beechcrafts at your fingertips.

"Being a Beech Aero Club member means you avoid any expense or responsibility of aircraft ownership. The club owns the airplanes, maintains them to perfection, pays taxes, hangar fees, and insurance.

"You just fly them . . . whenever you want.

"The Aero Club is a nationwide organization. Once a member, you are welcome at any of the Beech Aero Clubs across the U. S. where you can take advantage of their reduced rental rates and join in the social activities. Jackets, patches, and various gift items are available, too, and all members receive BAC TALK, the Beech Aero Club news. A National Aero Club Roundup is held yearly where members from all across the U. S. fly in for fun and activities.

"As a Beech Aero Club member, you'll have access to a complete range of exclusively-prepared training aids designed to help you advance your ratings and improve your skills. A Beech Aero

Club is also your local Flight Review Headquarters. Beech has produced three complete and unique training courses. The Private Pilot Course provides the student with all the knowledge needed to acquire his Private Certificate. The Beech Advanced Pilot Training Course is the fast efficient way to get your Commercial Certificate and/or Instrument Rating. All Beech training programs utilize the latest information, materials, and audiovisual techniques.

"In the Beech Flight Training Programs, the audiovisual portion of the lesson introduces the subject. Reinforcement follows with textbook, workbook and quizzes. This leaves the time completely open for instruction and practice. The final touch is the flight simulator which is worked into the curriculum to help them make the best of the student's valuable time.

"Flying can also be going somewhere and doing something fun. That's another area where the Beech Aero Club can assist you. The Aero Club manager or "club pro" will help you organize your flying activities. Functioning similar to a travel agent, he'll gather information and handle details, leaving you only the fun of it all.

"The Beech Aero Club offers you a sensible alternative to airplane ownership. Remember, your Beechcraft airplanes are ready and waiting at your local Beech Aero Club."

No, a commercial flying club isn't exactly aircraft ownership, but it's a start in that direction.

SOLE OWNERSHIP

Let's see now—work two jobs? Hope for an inheritance? Sell the house? Borrow on life insurance? Stay single? Find job for mate? Ask for raise? What other ways are there to afford sole ownership of that dream aircraft, at least for those born to the working class?

Impossible? No, just a bit more difficult. Now let's put things in order. A goal, any goal, must be obtainable in fact. So let's put the Learjet aside and settle for a new Cessna Cutlass. The goal is now reasonable but the down payment and monthly payments are still troublesome. Now we apply a bit of patience and begin the process of trading up to that realistic goal. The scenario goes like this:

Step 1. Purchase an older aircraft that has bottomed out in price but still has several hundred hours of engine life remaining. Be certain to purchase a common type of aircraft that will hold its resale value.

Step 2: Enjoy flying the machine while you pay off the loan. Absolutely *do not* add avionic equipment during your ownership period! At resale time you will lose most if not all of your investment in such equipment. Just keep what's there already in working order.

Step 3: Meticulously clean the aircraft, repaint if necessary, and sell or trade up to a larger used aircraft before the engine reaches TBO.

Step 4: Repeat steps 1 through 3.

Step 5: Now trade for that Cessna Cutlass.

It takes time to accomplish the process of acquiring an equity—several years, in fact. But the goal is eventually reached. What about that part-time job to expedite the process? Give it some serious thought. After all, almost anything beats watching TV nowadays.

Congratulations, aircraft owner! It's been a long haul but you made it.

Chapter 6
Aviation Career Opportunities

There are only two methods of personnel transportation in the United States—motor vehicle and aircraft. All other means—trains, ships, etc.—are trivial in terms of people carried. Highways and airways are the transportation paths for you and me. Of these two, the 55 mph speed limit has had an effect. As the time required to transit long distance by highway has been increased, more persons have taken to the air in spite of high fuel costs.

Aviation as an industry has grown steadily since the 1960s. The industry currently provides more than 400,000 jobs. Although the growth rate has stabilized in the past few years, it is predicted that the industry will support some 500,000 jobs by 1990. Aviation economic indicators continue to point upward for the long run—fast transportation of people and high-value cargo is fundamental to our way of life in the United States.

Yes, there are some slowing influences on the horizon. As a cost-saving technique, the two-pilot carrier aircraft is a strong reality. The increase in fuel costs is and will continue to cause second thoughts in the purchase of an airline ticket. Further, inflation has the same devastating effect on the airlines as it has on our personal paychecks. These adverse economic factors are undesirable. But strangely, the very same economic factors are spawning the development of a new generation of fuel-efficient air carrier aircraft, research on alternate fuels, and improved general aviation aircraft. NASA is beginning to reawaken to the needs of general aviation. The GAW and supercritical airfoils as well as winglets are examples of what research is providing to increase the efficiency of our general aviation aircraft. Further, experimentation with canard designs is providing improved levels of performance and, in some cases, stall-proof aircraft designs.

Growth in aviation will continue despite the current group of problem areas. Historically, aviation has always been faced with some form of economic problem. In time, each has been overcome. Just as the jet engine made air carrier transportation economical due to its excellent reliability and long life, the use of new composite aircraft materials, computer flight optimization techniques, and the like promise new economies tomorrow. In considering the overall situation, let us remember, in terms of real dollars, it cost more to learn to fly in the time of Lindbergh than it does today!

"Shall I make aviation my life's work? Beside the enjoyment of flying, what other benefits are there? What qualifications must I have? Is my pilot certificate all that is needed?" These and many other questions are logical to the young man or woman who is viewing aviation as a career. But there is one more question that is paramount to all of the others. Given the fact that you are healthy and of normal mental abilities, *are you willing to compete?* Since the beginning of flight, aviation has attracted the achiever, the student, the motivated, the person of action. The challenge of flight with its ever-present unknowns is the action an adventurous spirit craves. The high degree of technical sophistication in aviation is meat and potatoes to the engineering-minded. The perfectionist, too, has his place with precision flight and ATC procedures. The motivated can and always will be able to do whatever they set their minds on—a Mt. Everest is their opportunity, not their obstacle. Are *you* willing to compete? If so, join the thousands like you who are also interested in an aviation career and enjoy the spirited companionship as well as the exciting competition process.

INSTRUCTORS

The aviation flight training school is often the first step in the career of a commercial pilot. FAA-approved flight training schools are certified under FAR Part 141. This FAR establishes minimum limits for the classroom, the training equipment used, and the quality of instruction offered. Curriculum requirements as well as instructor requirements are specified for FAA-approved ground and flight schools. Aircraft maintenance facility requirements are additionally specified for flight schools. Fundamentally, the flight school offers aviation opportunities at three basic levels—the ground school instructor, the flight instructor, and the chief flight instructor. The following is a brief synopsis of minimum requirements for each of these levels.

Ground School Instructor

FAR Part 143 describes the basic requirements for a ground school instructor. To be eligible for a certificate, a person must be at least 18 years of age, of good moral character, and able to show his practical and theoretical knowledge of the subject for which he seeks a rating by passing a written test on that subject. Ground school instructor ratings are as follows:

Ground Instructor—Basic:	This rating qualifies the holder to instruct in a basic pilot ground school (private pilot).
Ground Instructor—Advanced:	This rating qualifies the holder to instruct in a basic or advanced pilot ground school (private and commercial pilots).
Ground Instructor—Instrument:	This rating qualifies the holder to give ground instruction in an instrument flying school, and for the operation of instrument procedures training devices.

Applicants who do not hold a ground instructor certificate must take a written test on fundamentals of instructing and a written test on the rating desired. Holders of a ground instructor certificate with one or more ratings are not required to retake the fundamentals of instructing test to obtain additional ratings. An applicant who holds a currently effective teacher certificate issued by a state, county, or city, authorizing him to instruct in a junior or senior high school, or who is regularly employed as an instructor in an accredited college or university, is not required to take the written test on the fundamentals of instructing.

The ground instructor rating is perhaps the simplest and most certainly the lowest-cost aviation certificate that is offered. The only real requirement is a bit of serious home study relating to the theory and practice of flight.

Ground school instruction opportunities exist at flight schools, junior colleges, community colleges, and, in some cases, at the university level. Earnings vary depending upon the course and level but are typically in the vicinity of $7.00 to $10.00 per hour. To retain currency, the holder of a ground instructor certificate must show that he has served for at least three months as a ground instructor out of the previous twelve months or demonstrate to the FAA that he meets the standards prescribed for the certificate and rating. The ground instructor rating is an excellent certificate to have, and one which many flight instructors acquire as a normal part of their activities. Although a pilot's certificate is not required to be a ground instructor, flight experience is certainly desirable.

Flight Instructor

The flight instructor is indeed a foundation stone of aviation. On his shoulders rests the quality and knowledgability of today's pilots. Most established flight training schools require that a flight instructor have a multi-engine and instrument ratings. Typical wages range from $8.00 to $10.00 per flight hour for part-time work to approximately $12,000 to $15,000 per year for full-time flight instruction work. There is little question that the flight instructor is an underpaid member of the aviation community. There are two basic reasons for this situation: First, the flight instructor job is often used by a new pilot as a means of acquiring experience in order to qualify for a more advanced flying position. Secondly, there are many who simply enjoy flying to the extent that they are willing to accept a low wage just to fly fine aircraft. One out of every 15 pilots is a certificated flight instructor.

Chief Flight Instructor

FAR Part 141 specifies that a chief flight instructor must be at least 21 years of age and have a good record as pilot and flight instructor. For a primary flying school, a chief flight instructor must have at least a commercial pilot's certificate with a flight instructor certificate and ratings for the category of aircraft to be used. In addition, at least 1000 hours pilot-in-command time is required. Primary flight instruction experience acquired as either a certificated flight instructor or an instructor in a military pilot primary flight training program is required to the extent of 1000 flight instruction hours or two years experience and a total of 500 flight instruction hours. Furthermore, within the year preceding designation as a chief flight instructor, an applicant must have given at least 100 hours of primary flight instruction as a certificated flight instructor in the category of aircraft used in the course.

For a commercial flying school or flight instructor school, a chief flight instructor must have at least a commercial pilot's certificate and a flight instructor's certificate, each with the rating for the category of aircraft to be used in the course. In addition, an instrument rating and 2000 hours as pilot-in-command are required. Flight instruction experience must consist of at least three years' experience and a total of 1000 flight hours or 1500 flight hours (in lieu of the three-year time requirement). Within the year preceding designation as a chief flight instructor, the applicant must have given at least 100 hours of pilot instruction as a certificated flight instructor in the category of aircraft to be used in the course. In addition, one year of active service as chief flight instructor of an approved primary flight course or one year of active service as an FAA-designated pilot examiner is required.

For an instrument flying school, a chief flight instructor must have at least a commercial pilot's certificate and a flight instructor certificate each with an instrument rating. In addition, 100 hours of flight time under actual or simulated instrument conditions are required and 1000 hours as pilot-in-command. Instrument flight instruction experience must consist of at least two years experience and a total of 250 flight hours (or 400 flight hours in lieu of the two-year time requirement). Within the year preceding designation as a chief flight instructor, the applicant must have given at least 100 hours of instrument flight instruction as a certificated instrument flight instructor or have acted in the capacity of an FAA designated instrument rating examiner during the course of the preceding year.

Wages for chief flight instructors vary significantly depending upon the size of the aviation flight school and the type of aircraft utilized. For a small flight school dealing principally in primary aviation, a typical wage for a chief flight instructor may be $15,000 per year. Chief flight instructors for large schools offering training directed toward commercial operations and heavy aircraft may command a salary of $25,000 per year.

AIR TAXI AND COMMERCIAL AIRCRAFT OPERATORS

Airborne operations relating to air taxis and commercial operation of aircraft have reached a high degree of sophistication during the past few years. The holder of an Air Taxi/Commercial Operator (ATCO) certificate is required to adhere to rigorous standards of operations, maintenance, and reporting. To qualify as pilot-in-command for VFR flights, an applicant must have at least 500 hours flight time as a pilot, including at least 100 hours of cross-country time. In addition, an instrument rating is required. For VFR flight at night, an additional requirement of 25 hours exists. To qualify as pilot-in-command for IFR flight, an applicant must have at least 1200 hours flight time as a pilot including 500 hours of cross-country flight time, 100 hours of night flight time, including at least ten night takeoffs and landings, and 75 hours of actual or simulated instrument flight time, at least 50 of which were in actual flight.

The pilot's proficiency requirements noted represent the *minimum* as specified by FAR 135. In practice, it is not unusual that air taxi operators require an Air Transport Pilot rating (ATP) to qualify for a pilot position. To retain proficiency as an ATCO pilot, FAR 135 requires that every VFR pilot-in-command be given a flight evaluation performance check at twelve month intervals; IFR pilots-in-command are required to be checked at six month intervals.

An applicant for the position of copilot (second-in-command) is required to have at least a current commercial pilot certificate with appropriate airplane category and class ratings for VFR flight. In the case of flights under IFR, a current instrument rating is additionally required.

FAR Part 61 requires that all ATCO pilots retain currency by virtue of recent flight experience. To carry passengers under VFR conditions, a pilot must have made at least three takeoffs and three landings to a full stop within the preceding 90 days in an aircraft of the same category, class, and type. IFR currency requires a minimum of 6 hours of instrument time under actual or simulated instrument conditions within the preceding six calendar months. Not more than 3 hours of synthetic instrument trainer time may be included in the 6 hours of required instrument time.

Typical pilot salaries as of 1979-1980 range as follows:

Copilots	$14,000 per year
Light twin pilots	21,000 per year
Light turboprop pilots	26,000 per year
Heavy turboprop pilots	31,000 per year
Jet pilots	38,000 per year
Chief pilots	60,000 per year

Agricultural Aircraft Pilots

Specialized aviation activities such as crop dusting are governed by specific federal aviation regulations designed to describe the special activity to be accomplished. FAR Part 137, Agricultural Aircraft Operation, governs the dispensing of poisons or other substances intended for plant nourishment, soil treatment, propagation of plant life, or pest control. To qualify as pilot-in-command, an applicant must hold a current commercial or airline transport pilot certificate and be properly rated for the aircraft used.

Unless an applicant has a record of proven safety in his flight operation and competence in dispensing agricultural materials or chemicals, he may be required to pass a test demonstrating his knowledge and skill to the chief supervisor of agricultural operation. The test of knowledge may consist of the following:

☐ Steps to be taken before starting operations, including a survey of the area to be worked.

☐ Safe handling of economic poisons and the proper disposal of used containers for such poisons.

☐ The general effects of economic poisons and agricultural chemicals on plants, animals, and persons, with emphasis on those normally used in the areas of intended operations, and the precautions to be observed in using poisons and chemicals.

☐ Preliminary symptoms of poisoning of persons from economic poisons, the appropriate emergency measures to be taken, and the location of poison control centers.

☐ Performance capabilities and operating limitations of the aircraft to be used.

☐ Safe flight and application procedures

In addition, a test of flying skill to demonstrate short field and soft field takeoffs, approaches to the working area, flare-outs, swath runs, pull-ups and turn-arounds may be additionally required. Agricultural pilot salaries range from $15,000 to $60,000 per year depending upon the crop, season, and location involved.

The field is difficult to enter from the standpoint of insurance-mandated experience requirements. Further, the field is declining in the need for pilots as newer, longer-acting plant chemicals are devised.

Flight Engineers

The first step toward becoming an airline transport pilot is the flight engineer's certificate. To be eligible for a flight engineer certificate, a person must be at least 21 years of age, able to read, speak and understand the English language, and hold at least a second-class medical certificate. Flight engineer's certificates are offered for reciprocating engine aircraft, turboprop-powered aircraft, and turbojet aircraft. An applicant for a flight engineer certificate must pass a written test on regulations concerning the duties of a flight engineer, the theory of flight, aerodynamics, meterology, and the various aspects of airplane equipment, systems, operating procedures, etc. Before taking the written test, an applicant must demonstrate that he has experience which will meet one of the following six criteria:

- ☐ At least three years of diversified practical experience in aircraft and aircraft engine maintenance (of which at least one year was in maintaining multi-engine aircraft), and at least 5 hours of flight training in the duties of a flight engineer.
- ☐ Graduation from at least a two-year specialized aeronautical training course in maintaining aircraft and aircraft engines (including six months maintaining multi-engine aircraft), and at least 5 hours of flight training in the duties of a flight engineer.
- ☐ A degree in aeronautical, electrical, or mechanical engineering from a recognized college, university, or engineering school; six months of practical experience in maintaining multiengine aircraft and at least 5 hours of flight training in the duties of a flight engineer.
- ☐ At least 200 hours of flight time in a transport category aircraft as pilot-in-command or second-in-command performing the functions of a pilot-in-command under the supervision of a pilot-in-command.
- ☐ At least 100 hours of flight time as a flight engineer.
- ☐ Successful completion of an approved flight engineer ground and flight course of instruction.

All references to flight experience refer to an aircraft that has at least three engines that are rated at least 800-horsepower each (or the equivalent in turbine engines).

In addition to the written examination, an applicant must pass a practical test on the duties of a flight engineer to demonstrate that he can satisfactorily perform preflight and inflight functions required for the position.

Typical airline standards for the position of flight engineer require that applicants be between 21 and 35 years of age, 5 feet 6 inches to 6 feet 4 inches in height, with at least two years of college training. A commercial pilot's certificate plus 500 to 1000 hours of flight experience may also be required. Salaries for the position of flight engineer typically begin at $18,000 per year. For additional information on certification requirements relating to flight engineers and also flight navigators, refer to FAR Part 63.

Airline Transport Pilot

Top of the line is the airline transport pilot. This lofty goal is reached only with a great deal of endeavor. FAR Part 61 describes the minimum requirements necessary to obtain an Air Line Transport Pilot certificate, the first major step to an airline pilot's career. As indicated by FAR, a candidate must be at least 23 years of age, of good moral character, able to read, write, and understand the English language, a high school graduate (or equivalent), and have a first-class medical certificate. In addition, an applicant must pass a comprehensive written examination on the fundamentals of air navigation, weather, regulations, radio communications, and other items pertaining to the airline transport field. In terms of experience, an applicant must have a commercial pilot certificate without limitations or must be a pilot in an armed force of the United States whose military experience qualifies him for a commercial pilot certificate. The applicant must have had at least 200 hours of flight time as pilot-in-command, 100 hours of which were cross-country and 25 hours which were night flights. In addition, 1500 hours of total flight time is required. Of this quantity, at least 500 hours must be cross-country flight time, 100 hours night flight time, and 75 hours of actual or simulated instrument time, at least 50 hours of which were actual. In certain instances, second-in-command time or flight engineer time may be credited against the requirement for 1500 hours of flight time. (See FAR Part 61 for details). A comprehensive oral examination and flight test is also required to obtain an Airline Transport Pilot certificate.

To reach the goal of airline transport pilot requires a great deal of skill and determination as well as the gift of good health and sufficient financial backing to acquire the minimum flight time experience. Employment as a flight instructor is one technique often used to obtain the necessary flight hours and skills. Flight training with the military is a second technique.

Current industry standards for airline transport pilot trainees are 20 to 35 years of age (21-28 preferred), 5 feet 6 inches to 6 feet 4 inches in height, 140 to 170 lbs., and at least two years of college (four years of college preferred). Copilot salaries average about $27,000 per year. Captain salaries average at approximately $60,000 per year. It is not unusual that a pilot will serve five to ten years as a flight engineer before being promoted to a copilot. Similarly, a copilot may serve in excess of ten years before becoming eligible for a captain position.

Flight Attendant

To qualify for an airline flight attendant position, an applicant must be at least 19 years old, in good health, and a high school graduate (two years of college preferred as a minimum). The flight attendant position does not require any form of flight experience, but does present an opportunity to be associated with the flying field. Salaries average from $10,000 to $15,000 per year. Opportunities for advancement in this field are quite limited.

AIRCRAFT MECHANIC

At the present time, the aviation industry, civil and military, employ about 135,000 aircraft mechanics. The profession is divided into three primary categories: airplane mechanic, powerplant mechanic, and aircraft inspector.

To qualify for either the airplane or powerplant mechanic certificate, an applicant must have at least 18 months of experience in the trade as well as pass both written and oral examination on the subject. In general, most mechanics obtain the airplane and the powerplant certificate for which they must have a total of 30 months of experience. In general, the experience requirement is satisfied by attending a trade school that provides a study program for the rating. Typically, such requires 18 to 24 months to complete.

The aircraft inspector certificate qualifies the holder to inspect the work of others and return an aircraft to service following its annual inspection or a major repair. A minimum of three years of experience as an airplane and powerplant mechanic is a necessary qualification, plus passing an FAA written and oral examination.

Salaries for aircraft mechanics range from $14,000 a year for a small shop to an average of approximately $27,000 per year for the airlines or a large repair facility.

The work is demanding in terms of precision craftsmanship and record keeping. Furthermore, evening or weekend work may be required in order that aircraft be made available to an operator during the normal work day. A pilot's certificate is not required for the occupation but is certainly an asset to understanding the mechanic's job. A mechanic with a pilot's certificate can test-fly aircraft for which he is rated. In this regard, the mechanic may directly evaluate an aircraft problem, or repair, and thus not be dependent upon the word of another pilot as to the performance of an aircraft.

AVIONICS REPAIRMAN

The avionics repairman is becoming more and more in demand as the electronic sophistication of aircraft communications, navigation, and autopilot equipment increases. The day of simple tube replacement is over. Today's *average* radios feature frequency synthesizers, digital filters, and other complex circuits. Similar equipment typical of the corporate aircraft employ microprocessors, digital memories, and related interface devices to perform the communications and navigation job. Current military aircraft are fully computerized in terms of their complement of avionics equipment.

To qualify as an FAA-rated avionics repairman, a candidate must show experience as a skilled electronic technician, plus specific training in the type of communications or avionics equipment for which the rating is sought. The latter is most often satisfied by attending training schools offered by the manufacturers of avionic equipment. An FCC radio operator's license is also required by most avionics repair shops. Shortages currently exist in the fields of skilled radar and autopilot repairmen.

A pilot's certificate is not required to become an avionics technician; however, just as the aircraft mechanic, a pilot/repairman can often best evaluate a problem or repair by flying the aircraft. This is particularly important from the standpoint of autopilot installation and repair. Salaries for avionics repairmen are on a par with the aircraft mechanic.

In general, the avionics repairmen are associated with a fixed base operator or an aircraft repair facility. This is necessary in that the installation of avionics equipment often requires cutting holes in the skin of aircraft, modifying aircraft for wiring, etc. Since the aircraft structure is involved, the modification is usually accomplished (or supervised) by a mechanic with an airplane rating. Depending upon the extent of the modification, the mechanic may additionally be required to provide a new weight and balance calculation for the aircraft.

The work involved with avionics installation and repair is

demanding both mentally and physically. Electronic "bugs" are far more vexing to find than their mechanical counterparts. Furthermore, there is just nothing glamorous about crawling under an aircraft instrument panel when the plane has been sitting in a 110° sun all day.

Another occupation similar to the avionics repairman is the instrument repairman. This occupation also requires specialized training.

AIR TRAFFIC CONTROL

Our system of skyways is greatly dependent upon the "flying skill" of a vast armada of tower operators, radar operators, and flight service personnel. The term "flying skill" is not without basis, for today more than 50 percent of our air traffic control operators and flight service station specialists have received pilot training of one form or another in the course of their work. There are currently approximately 25,000 controllers employed by the Government.

Air Traffic Controllers work in three basic specialties: towers, en route centers, and Flight Service Stations.

Towers—In operating towers, Air Traffic Controllers control flights within the three- to thirty-mile radius each serves, under visual flight rules (VFR) or instrument flight rules (IFR). Tower controllers may work either in the glass-walled room at the top of the tower or in the radar room below it, but their jobs have the same aim—the safe separation and movement of planes within their area in taking off, landing, and maneuvering.

En Route Centers—These controllers work in centers control traffic operating along established airways across the country between tower jurisdictions. They maintain a progressive check on aircraft, issuing instructions, clearances and advice, as well as initiating search and rescue operations to locate overdue aircraft.

Flight Service Stations—Flight Service Stations provide assistance to pilots who must obtain information on the station's particular area, including terrain, weather peculiarities, preflight and in-flight weather information, suggested routes, altitudes, indications of turbulence, icing, and any other information important to the safety of a flight.

Virtually all controller jobs involve shift work because most facilities operate on a 24-hour basis. The exact rotation of the shift is usually determined by the individual facility but typically it might involve working several weeks from midnight until 8 a.m. followed by several weeks working from 4 p.m. until midnight; days off do not always fall on weekends.

Generally, training combines study at the FAA Academy in Oklahoma City with training at the facility where the recruit is assigned, although the order of the sessions is flexible. Facility training varies somewhat, but is a combination of on-the-job training and classroom time which continues until the controller reaches the journeyman level for the position assigned.

Air traffic control specialists who fail to maintain standards during training may be dropped from the program, and all controllers are required to meet qualification requirements and semiannual proficiency checks. Controllers who transfer from one facility to another with a higher traffic density must meet the checkout standards of the new facility, and those who remain in one facility must demonstrate their continued ability to handle the assignment. Failure to do so may result in demotion, separation, or reassignment to another facility where the load is lighter.

Journeyman level varies from facility to facility, depending on the type of assignment and the density of traffic; at towers and centers it is generally GS-13 or -14 and at Flight Service Stations, it is GS-10 or -11. Although promotions are not automatic, but depend on developing proficiency within prescribed time periods, most controllers reach journeyman level in three or four years.

OTHER AVIATION CAREERS

The aviation field is indeed large and embraces many technical disciplines. Just to name a few more, one might consider the following:

Military—All three military services offer superb aviation training in terms of pilots as well as repairmen. It has been estimated that the value of the training provided a graduate of the Air Force Academy is in excess of $250,000. In reality, one of the biggest sources of trained pilots employed by the airlines is from the military. The Army is perhaps the major supplier of helicopter pilots for civil jobs.

Administrative—The field of airport management as well as both state and Federal Departments of Transportation require skilled aviation administrative services to perform their function. This new and developing field should not be overlooked.

Engineering—Aviation and engineering go hand in hand. The aeronautical, mechanical, electrical, and chemical engineer all contribute to aviation as well as the mathematician and the physicist.

Salaries in this field are excellent, beginning at about $18,000 a year and averaging about $40,000 per year for a senior engineer.

Sales—Aircraft selling is indeed a specialty. A qualified salesman is really a tax and business specialist. It is up to the salesman to show how the aircraft will provide a return on the investment of the buyer. Except for aircraft sold purely for sport flying, the average buyer is interested in an aircraft as a business asset.

Service—Aircraft parts handling both wholesale and retail is another specialized service. Aircraft parts are not only high-value items, they must be delivered quickly. A grounded aircraft awaiting parts is a significant business liability.

Specialty pilot services—As an example of specialty pilot services the fields of TV, police, fire, forest patrol, and the border patrol require pilots to assist in performing the service at hand. In general, pilots who fulfill these roles have knowledge or experience in the basic service being offered. Stated another way, a police helicopter pilot, for example, is generally a commissioned policeman; likewise with the border patrol pilot, etc.

The list of aviation occupations continues—the fixed base operator, the aviation lawyer, writer, aerial photographer, borate bomber pilot. The aviation teacher, flight controller, design engineer, and administrator—all contribute to flight. The wings of aviation are indeed broad, providing for an army of design and support non-flying personnel without which aviation, as we know it today, could not exist. And, in fact, a very large percentage of the non-flying aviation support careers pay far better than their flying counterparts.

Seen from "above," there is a common thread that runs through both the flying and non-flying aviation community—the ability to successfully compete. The mystique of aviation continues to summon the aggressive as it has since Lilienthal proved that a heavier-than-air machine was practical. To the candidate who is willing to undertake a serious program of study and preparation, the opportunities of aviation are available. To those who have not reached this level of commitment, it would probably be best to look into some other occupational field.

Aviation will continue to become more technical, more specialized, more demanding—progress and growth are inevitable. Indeed, it often appears the day is not far away when the pilot certificate will warrant a college degree. If technical growth and new challenges are to your liking, come on in—the weather's fine and there is far to go.

Chapter 7
Problem Areas

Surprisingly, the announcement "I'm going to learn to fly" is sometimes greeted with less than an enthusiastic response. For some strange reason, it runs a close parallel to the statement, "I'm going to buy an airplane." It's really difficult to imagine why, but replies may be heard such as: "You? Oh, no!" or "Fly? It's not safe!" or "You're too old!" or "It's too expensive!" or "What for?" or "You haven't time," and on and on . . .

No, not everyone is possessed with the same enthusiasm for flying that posesses you. Your goals are foreign; they are not understood nor accepted by others. A problem exists—you have defied convention!

The decision to fly is significant. Initial problem areas are not uncommon. Let's look at a few of the classic obstacles to flight and also some possible solutions.

LACK OF ACCEPTANCE

Perhaps the most common problem of all is lack of acceptance to flying by one's mate. The syndrome is not reserved to wives; I have seen husbands violently opposed to wives taking to the air. What then is the basic problem?

In general, the root reason for lack of acceptance to flying is lack of understanding of the flying art. It is not unusual for one to truly fear for the safety of his or her partner and, hence, express displeasure at the thought of one's learning to fly. To one whose only education in flight has been newspaper headlines, the thought of flying can be ominous indeed.

This situation is so common that most flight schools make provisions to accommodate man/wife teams even though only one is actively taking flight training. A progressive ground school will allow the partner to sit in on lectures (at no extra charge) in recognition of the fact that the student will always do better when flying is accepted on a family basis. By sitting in on lecture sessions, "newspaper knowledge" is eventually replaced by understanding. Furthermore, a common core experience develops between partners as learning takes place. Advancing one step further, many flight schools offer copilot courses. Again, the theme is one of team involvement and mutual understanding of the art.

To be accompanied by a competent partner and copilot is a genuine pleasure, particularly when the weather becomes marginal or a fast-food lunch suddenly becomes rebellious. Absolutely nothing beats a well-coordinated team—flying safety, flying fun, and sharing new experiences are all enhanced. The key to acceptance is *understanding and participation.*

Regrettably, but factually, there exists a second motive for lack of acceptance to flying—namely, competing reasons for the use of funds. Road blocks by the score can be thrown up on this one. Considering first that the family's physical, medical, and safety needs have been satisfied, just how does the obtainment of a new skill trade off with a new car, new furniture, sending the daughter to an out-of-state college, or a vacation in Hawaii? You, of course, are the only one who can make the tradeoff. Is the old furniture really that uncomfortable, the existing car that unrepairable, the out-of-state college that beneficial, or Hawaii that pleasurable? Probing even more deeply, who will make the final tradeoff analysis?

Flying demands the ability to make decisions. You and you alone must equate the progress of your course against the whims of the weather. Weathering a financial storm requires the same kind of decision making—a cold, concise assessment of the facts. If the dollars simply aren't there, plan a way to make them available at a not-to-distant future date. If the dollars *are* there, in fact, and it's your turn, take it! Chart your financial flying course and go!

Problems with the partner accepting the compromise (the new furniture lost out)—a cumulonimbus cloud inside the house? A deep cold front in the living room in spite of the thermostat setting? Well, even the most severe of storms subside in time. When appropriate, try a little partner involvement in your flying program. Surprisingly, it works quite well.

In the final analysis, attainment of your goals are beneficial to both yourself and your partner.

Regrettably, there are cases where acceptance of one's flying goals is never obtained. The root cause may be fear of flight, lack of understanding, or disinterest. Flying now becomes a solo activity. The partner will simply never know the thrill of a dawn takeoff, the beauty of viewing a rainbow from above, the intricate structure of the clouds, or the magnificance of night flight. The adventure must be undertaken solo.

TIME

Just add water and boil for 60 seconds . . . Lose six pounds in six days . . . No experience necessary, make $2000 a month by simply . . . Nothing down, just walk in and drive out in a new . . . Learn to speak French in just four weeks . . . Buy now and save . . . Pass the FAA written in only one weekend. Instant response, instant achievement, instant success! Verily, we are in the *now* generation.

What a rude discovery, the realization that ground school requires homework—understanding weather requires study—radio navigation requires practice—*learning to fly takes time!* Solo in just ten hours, buy a plane, fly to Bermuda—now you tell me it will take *eight months* just to get a private pilot certificate; an instrument rating another eight months or more?

Instant success is a true problem of our times. So often the beginning student is dismayed to find that the private certificate will require a good outlay of time as well as money. So accustomed have we become to the teachings of TV that we just naturally apply them to any new endeavor. Learning to fly is just one example.

Time is a necessary ingredient to the flying process. Put aside the TV, cancel the newspaper subscription, let the world wait. You have decided to learn to fly. Make the time needed and enjoy every moment of the learning process. As has been stated, "Flying itself is not inherently dangerous, but it is terribly unforgiving of carelessness or mistakes." For this reason, if no other, take the time required to learn the art well.

As our society has continued to grow in complexity, so has the art of flying. Just within the past ten years or so the amount of flying knowledge required of a student has doubled. Today's private pilot must be familiar with radar navigation, TCAs, TRSAs, operations at

117

busy air carrier airports, wake turbulence effects from heavy aircraft, and flight in crowded airways. Today's private pilot is as proficient as yesterday's commercial aviator. In tomorrow's world it may well be that a college degree will be required for a pilot certificate.

It takes time to achieve the needed level of proficiency. There is no such thing as instant learning. If for no other reason than to get your money's worth, plan to take the time to get the very most out of *every* learning session, every flight. Ask your flight instructor to recommend reading material to supplement your formal study material. If possible, join AOPA* and keep current on what's going on in today's world of flight. For the aviation businessman, the publication *Business and Commercial Aviation*** may be beneficial.

Whatever the case, commit! Become totally involved! Take the time needed to do the learning job well. Turn the TV to the wall and fly.

MATURITY

Flying requires maturity. The flight decisions you make directly affect your well-being as well as the health and safety of your passengers. Decision making, self discipline, and flying go hand in hand. Maturity is the key to true mastery of flight.

Jim R. attended class regularly but his disinterest in the subject was most evident. Slides and movies as visual aids were welcomed as interludes for sleep. Quizzes were finished quickly, followed by a rapid departure from the classroom. Jim's final exam netted a grade of 32. Exams do tell a story. In this instance, the entire ground school session told a story. Jim R. was 17 years of age, the son of a very successful airline pilot. Father had decreed that son would follow in the profession—so off to ground school. Jim, exhibiting the classic symptoms of our time, demonstrated total immaturity, by his rebellious and indifferent approach to class. No, Dad, not yet—give son a few years to find maturity, *then* introduce him to flying! A student learns only when he or she is ready to learn, and not before. Flying will not create maturity, maturity must come first.

Pat B. soloed on her 20th hour of flight instruction. After an "almost" accident while on her third solo flight, she sought out a second flight instructor (me, unfortunately) for an opinion of her airwork. The experience was unforgettable—absolutely no regard for airspeed, altitude, the key position, and other elements of a

*Aircraft Owners and Pilots Association. P.O. Box 5800, Washington, D.C. 20014
**Ziff-Davis Publishing Co, One Park Avenue, New York, New York 10016

landing. Each entry to the airport was different from the preceding in one element or another. Each flare and touch approached uncontrolled flight as the plane was herded onto the runway. Two such sessions added several gray hairs to my head but little to Pat's flying skill. She was a bright girl, very quick and well-coordinated, but lacking in discipline and devoted to that popular college creed, "Do your own thing—your own way." Pat B., age 26—smart, educated, dynamic, and very immature. No, Pat; flying is not for you, at least not yet. Flying is discipline—self-discipline as well as conformity to the flight practices and procedures that grant longevity to the pilot. Come back in a few years, Pat, when your devotion to the granny glasses and tattered jeans has dimmed. Come back when your desire to learn and understand exceeds your desire to be recognized. Experiences more significant than simple nonconformity await you. Yes, Pat, you have all the abilities to fly except one: maturity—acceptance of discipline for those arts that require discipline. Flying is not for you, Pat—not yet.

FEAR

A classic problem leading to temporary fright rests with the overzealous student on his or her first few solo flights. The scenario often begins with a full power stall, a very high nose attitude, the rapid drop of one wing, and then an excellent view of the earth—straight down. Chop power, relax back pressure, and somehow the aircraft finally recovers. That's enough for today; let's get this thing back on the ground and think about this flying business. Maybe it's not so great after all. "Anyhow, why do I need to do stalls? I just wanted to learn to fly, not all this other stuff."

To the over-ambitious student who is going to "see what this airplane can do," an early fright is simply a lesson in humility. Suffice it to say he may next ponder the statement, "There are old pilots and there are bold pilots, but there are no old, bold pilots."

To the student who is truly surprised by an unexpected flying event, look critically at your flight instructor. You have not been adequately prepared. The Federal Aviation Regulations, as they pertain to student training, and the flight training curricula employed today are designed to provide adequate pre-solo experience such that a student can perform assigned exercises without difficulty—some apprehension perhaps, but true fright, no. Your instructor should demonstrate to you how the airplane responds when a maneuver is done *correctly* and how it responds when done *incorrectly*. In either event, the instructor must assure both yourself

119

and himself that you are fully capable of recovery procedures. Only then is solo practice to be authorized.

A third possibility exists in the event you have chosen the lowest possible cost flight school or instructor. Here, dollar compromises have been made and one of them may be in the adequacy of your preparation. Examine the private pilot curriculum contained in the next chapter of this text. Have you had *all* of the flight operations listed therein? Has your instructor demonstrated spin recovery? If either of these questions render a no answer, better have a talk with your instructor. Granted, the FARs do not currently *require* the teaching of spin recovery, but student pilots can make errors during the practice of stalls. The time to learn spin recovery is in pre-solo dual flight training, not during an actual spin while solo.

There is nothing basically wrong with attending a low-cost training facility (so long as it meets FAA requiements). Simply remember that a cost compromise exists. You may need to take the responsibility of assuring yourself that you are being given an adequate minimum of training. It may well be up to you to *ask* for another dual session, or to repeat an exercise, etc., until you feel confident of the maneuver or exercise.

Other moments of apprehension during training: The aircraft's only radio quits while entering a busy airport. Checkpoints don't appear to be working out on a cross-country; the terrain somehow just doesn't appear right. Rain? That certainly wasn't planned. Neither was the cloud buildup. Yes, flying does and always will have its moments of apprehension. They come less often with experience but are still there.

In closing this short note addressed to a perfectly natural response—fear—simply remember it is common to *all*. Charlie C., a flying associate of mine, can only be described by the word "tough"—a hard-driving businessman in a tough business in a tough neighborhood; he's a strong tennis player and a fine VFR pilot. But put Charlie under the hood and bank the aircraft 60 degrees and we have pussycat Charlie, guaranteed every time. Given the right stimuli, even the bravest exhibit fear.

Contained within the flying art is such an abundance of experiences that an occasional humbling incident becomes inevitable to all. Weather is, without doubt, the greatest teacher of humility. Expect an anxious moment every now and then. It's simply a part of flying. If you have done a complete job during your training, you have with you the tools to handle such an event; it becomes simply an incident in your memories of flight.

Chapter 8
A Private Pilot Training Program

To derive the maximum benefit from a private pilot training program, there is much the student should know in advance about the course he or she is planning to take. "What are the contents of the course; how many lessons are there? How much time will be needed for homework? Is there anything that can be done to minimize costs?" Questions such as these are rightfully those of the student and should be answered in full *before* beginning a course of instruction.

In modern teaching it is absolutely proper to provide the student a complete course outline or curricula at the first class meeting. As noted in Chapter 3, a student will always do better if provided a clear understanding of where the course is going and where he or she stands at the moment. This principle is paramount among the principles of teaching! It is for this reason that this chapter and the next have been included in this text.

USING A CURRICULUM

The training program curriculum presented in this chapter is devoted entirely to the private pilot who is seeking his or her single engine landplane certificate. For convenience in reading it is broken into the following parts:

Part 1: A ground school curriculum.
Part 2: A flight instruction curriculum.
Part 3: The private pilot flight test.

Part 1 outlines material that is to be covered in a private ground school. Since the Federal Aviation Regulations provide for both approved ground school courses and home study programs*, Part 1 may be used as a guide to designing an acceptable home study program. Part 2 describes a typical flight instruction curricula for the single engine landplane pilot certificate. In reality, the last lesson of the flight instruction course is the FAA flight test. By knowing exactly what the test consists of and the criteria for acceptable performance, the student is armed with the knowledge he or she will need to assure a successful outcome. Part 3 describes the flight test process.

Parts 1 and 2 are extracted from the *Aviation Instructor's Handbook,* AC 60-14 as published by the Department of Transportation, Federal Aviation Administration. Similarly, Part 3 is extracted from the *Private Pilot-Airplane Flight Test Guide,* AC 61-54A. Where appropriate, minor editing has been performed to enhance clarity. To obtain the maximum benefit from Parts 1 and 2, check-off each item in a block of learning as the material is presented. This will aid in assuring that the instruction content is complete from the standpoint of a basic course.

A curriculum such as the one presented here cannot be other than general in content. It is up to the flight/ground school instructor to add those blocks of learning peculiar to your location or planned flight routes. For those located in the Northern parts of the United States, additional emphasis must be placed on cold weather flying; for Western area pilots, mountain flying techniques should be added; for the Gulf and Coastal regions, supplemental flight training relating to thunderstorms and fog is needed. Similarly, pilots operating in areas of heavy traffic must insist that their curriculum be modified to contain in-depth training relating to TRSAs (Terminal Radar Service Areas) and TCAs (Terminal Control Areas).

Any practical training syllabus must be flexible, and should be used primarily as a guide. The order of training can and should be altered, when necessary, to suit the progress of the student and the demands of special circumstances. In departing from the order prescribed by the syllabus, however, it is the responsibility of the instructor to consider the relationships of the blocks of learning affected. It is often preferable to skip to a completely different part of the syllabus when the conduct of a scheduled lesson is impossible, rather than proceeding to the next block which may be predi-

---

*FAR 61.35 (a) (1)

cated completely on skills to be developed during the lesson which is being postponed.

In closing this introduction, it is recognized that lists of training material and examination criteria in themselves do not make exciting reading. Their inclusion is not so much for reading as for a checklist to weigh your training against. Are you getting your money's worth? Is your training program complete? Each lesson of the private pilot training program which follows sets forth a unit of instruction. Neither the time nor the number of school periods to be devoted to each lesson is specified. The sequence in which the sample lessons are listed is not necessarily the most desirable one to use in all training situations and should be varied as necessary to suit the circumstances.

PART 1: A GROUND SCHOOL CURRICULUM

Refer to the Pilot/Controller Glossary, at the back of this book for a definition of abbreviations used in this chapter.

Lesson No. 1

Objective. To develop the student's knowledge with regard to the definitions and abbreviations in Part 1 and the appropriate regulatory requirements of Part 61 of the Federal Aviation Regulations.

Content.
1. Airplane Registration and Airworthiness Certificates.
2. FAR, Part 1—Definitions and abbreviations important to a private pilot.
3. FAR, Part 61.
   a. Requirements for certificates and ratings.
   b. Duration of pilot certificates.
   c. Medical certificate requirements.
   d. Written tests.
   e. Flight tests.
   f. Pilot logbooks.
   g. Recency of experience (including biennial flight review).
   h. Private pilot privileges and limitations.

Completion Standards. The lesson will have been successfully completed when, by an oral test, the student displays a working knowledge of the appropriate portions of FAR Part 1 and Part 61, and demonstrates the ability to locate and use information in these rules.

Lesson No. 2

Objective. To develop the student's knowledge of the pertinent regulatory requirements of Part 91 of the Federal Aviation Regulations and the accident reporting rules of the National Transportation Safety Board as they relate to private pilot operations.

Content.

1. FAR, Part 91.
   a. General operating and flight rules.
   b. VFR requirements.
   c. IFR requirements (familiarization).
   d. Maintenance, preventive maintenance, and alterations.
   e. Familiarization with Subpart D.
2. National Transportation Safety Board Procedural Regulations, Part 830—Notification and Reporting of Accidents.

Completion Standards. The lesson will have been successfully completed when, by an oral test, the student demonstrates the ability to locate and use information in the appropriate rule as related to private pilot operations.

Lesson No. 3

Objective. To develop the student's knowledge of the Airman's Information Manual as it relates to VFR operations and to develop competence in using the Advisory Circular System.

Content.

1. Airman's Information Manual as it relates to:
   a. Air Navigation Radio Aids.
   b. Airports and Air Navigation Lighting and Marking aids.
   c. Airspace.
   d. Air Traffic Control.
   e. Services Available to Pilots.
   f. Airport Operations.
   g. Emergency Procedures.
   h. Good Operating Practices.
   i. Airport Directory (legend).
   j. Airport Facility Directory (legend).
   k. Graphic Notices and Supplemental Data.
2. FAA Advisory Circular System—Series 00, 20, 60, 70, 90, 150, and 170 (familiarization).

Completion Standards. The lesson will have been successfully completed when, by an oral test and demonstration, the student displays a basic knowledge of appropriate Parts of the Airman's Information Manual and the FAA Advisory Circular System.

Lesson No. 4

Objective. To develop the student's knowledge of the operation of aircraft radios, the use of proper radio phraseology with respect to air traffic control facilities, and to develop competence in the use of the slide rule face of the flight computer and aeronautical charts in planning a VFR cross-country flight.

Content.

1. Radio communications.
   a. Operation of radio communications equipment.
   b. Ground control.
   c. Tower.
   d. ATIS.
   e. Flight service station.
   f. UNICOM.
   g. Technique and phraseology.
2. ATC light signals.
3. Flight computer—slide rule face.
   a. Time.
   b. Speed.
   c. Distance.
   d. Fuel consumption.
4. VFR navigation.
   a. Aeronautical charts.
   b. Measurement of courses.
   c. Pilotage.
   d. Dead reckoning.

Completion Standards. The lesson will have been successfully completed when, by an oral test and demonstration, the student displays a basic knowledge of radio communications, ATC facilities, and aeronautical charts, and is able to use the flight computer to solve elementary VFR navigation problems.

Lesson No. 5

Objective. To further develop the student's knowledge of pilotage, dead reckoning, and radio navigation.
Content.

1. VFR navigation.
   a. Pilotage.

    b. Dead reckoning.
2. Operation of the navigational radio equipment.
    a. VOR.
    b. ADF.
    c. Use of radio aids.
3. Flight computer—wind face.
    a. Determination of wind correction angle and true heading.
    b. Determination of groundspeed.
4. Flight computer—slide rule face.
    a. Review time, speed, and distance problems.
    b. Review fuel consumption problems.

Completion Standards. The lesson will have been successfully completed when, by an oral test and demonstration, the student displays a basic knowledge of VFR navigation and the use of radio aids. The student should be able to solve fundamental and advanced problems on the flight computer.

Lesson No. 6

Objective. To review Lesson 5 and thereby improve the student's competence in VFR navigation procedures; to introduce advanced VFR radio navigational problems; to develop the student's knowledge of emergency procedures with respect to VFR cross-country flying; and to introduce flight planning.

Content.
1. Review of Lesson 5.
2. Use of ADF.
3. Radar.
4. Use of VOR, intercepting and maintaining radials.
5. Emergency procedures.
    a. Diversion to an alternate.
    b. Lost procedures, including the use of radar and DF instructions.
    c. Inflight emergencies, including emergency landings.
6. Transponder.
7. DME.
8. Flight planning.

Completion Standards. The lesson will have been successfully completed when, by an oral test and demonstration, the student displays a working knowledge of advanced VFR radio navigational procedures, cross-country emergency procedures, and can accurately plan and plot a VFR cross-country flight.

Lesson No. 7

Objective. To further develop the student's competence in flight planning and to acquaint the student with the medical factors related to flight and general safety precautions.

Content.
1. Flight planning.
2. Medical factors related to flight.
   a. Fatigue.
   b. Hypoxia.
   c. Hyperventilation.
   d. Alcohol.
   e. Drugs.
   f. Vertigo.
   g. Carbon monoxide.
3. General safety.
   a. Ground handling of aircraft.
   b. Fire—on the ground and in the air.
   c. Collision avoidance precautions.
   d. Wake turbulence avoidance.

Completion Standards. The lesson will have been successfully completed when, by an oral test, the student displays a basic knowledge of flight planning, the medical factors related to flight, and general safety procedures.

Lesson No. 8

Objective. To develop the student's knowledge of the fundamentals of weather, as associated with the operation of aircraft.

Content.
1. Atmospheric layers.
2. Pressure.
3. Circulation.
4. Temperature and moisture.
5. Stability and lapse rate.
6. Turbulence.
7. Clouds.
8. Airmasses.
9. Fronts.
10. Aircraft icing.
11. Thunderstorms.

Completion Standards. The lesson will have been successfully completed when, by an oral test, the student demonstrates a fundamental knowledge of aviation weather.

Lesson No. 9

Objective. To develop the student's ability to interpret and use weather charts, reports, forecasts, and broadcasts; and to develop the student's knowledge of the procedure for obtaining weather briefings.

Content.
1. Review Lesson 8.
2. Weather charts.
   a. Weather depiction charts.
   b. Surface prognostic charts.
3. Aviation weather reports.
   a. Hourly sequence reports.
   b. Special surface reports.
   c. Pilot reports.
   d. Radar reports.
4. Aviation weather broadcasts.
   a. Transcribed weather broadcasts.
   b. Inflight weather advisories.
5. Weather briefings.
6. Review requirements of regulations for VFR flight.
7. Aviation weather forecasts.
   a. Area forecasts
   b. Terminal forecasts.
   c. Winds aloft forecasts and reports.
   d. Route forecasts.

Completion Standards. The lesson will have been successfully completed when, by an oral test and demonstration, the student displays the ability to interpret and use weather charts, reports, forecasts, and broadcasts, and can obtain and understand a weather briefing.

Lesson No. 10

Objective. To further develop the student's knowledge of aviation weather through a review of Lessons 8 and 9; to develop the student's knowledge of Greenwich time; and to develop the student's ability to recognize various weather conditions.

Content.
1. Review of Lessons 8 and 9.
2. Greenwich time.
3. Weather recognition.

Completion Standards. The lesson will have been successfully completed when, by an oral test, the student displays a working knowledge of Greenwich time, and a knowledge of how critical weather situations can be recognized both from the ground and during flight.

Lesson No. 11

Objective. To develop the student's knowledge of airplane structures, propellers, engines, systems, and the magnetic compass.

Content.
1. Airplane structures.
   a. Construction features.
   b. Flight control systems.
   c. Rigging.
2. Propellers.
   a. Fixed pitch.
   b. Controllable.
3. Reciprocating airplane engines.
   a. Construction features.
   b. Principle of operation—four stroke cycle.
   c. Fuel system, including carburetors and fuel injectors.
   d. Lubrication system.
   e. Ignition system.
   f. Engine instruments.
   g. Operating limitations.
   h. Malfunctions and remedial actions.
4. Airplane hydraulic system.
   a. Principle of hydraulics.
   b. Use of hydraulics in airplanes.
   c. Construction features of a simple airplane hydraulic system.
   d. Retractable landing gear and flaps.
   e. Malfunctions and remedial actions.
5. Airplane electrical system.
   a. Fundamentals of electricity.
   b. Operation of airplane electrical power system units.
   c. Electrically operated flight instruments.
   d. Retractable landing gear.
   e. Flaps.
   f. Fuses and circuit breakers.
   g. Malfunctions and remedial actions.
6. Pitot-static system and instruments.
   a. Airspeed indicator, including markings.

b. Altimeter.
   c. Vertical-speed indicator.
7. Vacuum system and instruments.
   a. Attitude indicator.
   b. Heading indicator.
   c. Turn and slip indicator.
8. Magnetic compass.
   a. Errors.
   b. Use in flight.

Completion Standards. The lesson will have been successfully completed when, by an oral test, the student displays a basic understanding of airplane structures, engines, systems, and instruments.

Lesson No. 12

Objective. To develop the student's knowledge of basic aerodynamics.

Content.
1. Forces acting on an airplane in flight.
   a. Lift.
   b. Weight.
   c. Thrust.
   d. Drag.
2. Airfoils.
   a. Angle of incidence.
   b. Angle of attack.
   c. Bernoulli's Principle.
   d. Newton's Laws.
3. Factors affecting lift and drag.
   a. Wing area.
   b. Airfoil shape.
   c. Angle of attack.
   d. Airspeed.
   e. Air density.
4. Function of the controls.
   a. Axes of rotation—longitudinal, lateral, and vertical.
   b. Primary controls—ailerons, elevators, and rudder.
   c. Secondary controls—trim tabs.
   d. Flaps and other high-lift devices.
5. Stability.
   a. Static stability.
   b. Dynamic stability.

6. Loads and load factors.
    a. Effect of bank angle on stall speed.
    b. Effect of turbulence on load factor.
    c. Effect of speed on load factor.
    d. Effect of load factor on stall speed.
7. Torque.
    a. Gyroscopic reaction.
    b. Asymmetrical loading to propeller ("P" factor).
    c. Slipstream rotation.
    d. Torque reaction.

Completion Standards. The lesson will have been successfully completed when, by an oral test, the student displays an understanding of basic aerodynamics.

Lesson No. 13

Objective. To develop the student's knowledge of the fundamental flight maneuvers.

Content.
1. Straight-and-level flight.
    a. Pitch, bank, and yaw.
    b. Trim.
    c. Integrated use of outside references and flight instruments.
2. Level turns.
    a. Forces acting in a turn.
    b. Aileron drag and coordination.
    c. Speed of roll.
    d. Slips and skids.
    e. Integrated use of outside references and flight instruments.
3. Climbs and climbing turns.
    a. Best rate-of-climb airspeed.
    b. Best angle-of-climb airspeed.
    c. Torque and coordination.
    d. Trim.
4. Glides and gliding turns.
    a. Effect of high lift devices.
    b. Most efficient glide speed.
    c. Coordination.
    d. Trim.
5. Descents with power.
    a. Power settings and airspeeds.
    b. Trim.

Completion Standards. The lesson will have been successfully completed when, by an oral test, the student displays a basic understanding of the fundamental flight maneuvers.

Lesson No. 14

Objective. To develop the student's ability to properly use Pilot's Operating Handbooks and FAA Approved Airplane Flight Manuals; to develop the student's ability to perform basic weight and balance computations; and to develop the student's understanding of fundamental flight training maneuvers.

Content.
1. Use of data in Pilot's Operating Handbook or FAA Approved Airplane Flight Manual.
   a. Takeoff and landing distances.
   b. Fuel consumption and related charts.
   c. Maximum range power settings.
   d. Maximum endurance power settings.
2. Weight and balance.
   a. Terms and definitions.
   b. Effects of adverse balance.
   c. Finding loaded weight.
   d. Finding center of gravity—when weight is added or removed—when weight is shifted.
3. Maneuvering at minimum controllable airspeed.
4. Stalls.
   a. Theory of stalls.
   b. Imminent stalls—power-on and power-off.
   c. Full stalls—power-on and power off.
5. Flight maneuvering by reference to ground objects.
   a. S-turns across a road.
   b. Rectangular course.
   c. Eights along a road.
   d. Eights across a road.
   e. Turns around a point.
   f. Eights around pylons.

Completion Standards. The lesson will have been successfully completed when, by an oral test and demonstration, the student displays a basic, knowledge of Pilot's Operating Handbooks and FAA Approved Airplane Flight Manuals; when the student is able to perform basic weight and balance computations; and when the student has a working knowledge of the performance of fundamental flight training maneuvers.

Lesson No. 15

Objective. To develop the student's knowledge of fundamental flight maneuvers and attitude instrument flying.

Content.

1. Review Lesson 14.
2. Takeoffs and landings.
   a. Normal and crosswind takeoffs and landings.
   b. Soft field takeoffs and landings.
   c. Short field takeoffs and landings.
   d. Go-arounds or rejected landings.
3. Introduction to attitude instrument flying. Maneuvering by reference to flight instruments—pitch, power, bank, and trim control in the performance of basic flight maneuvers.
   a. Straight-and-level flight.
   b. Turns.
   c. Climbs.
   d. Descents.
   e. Recovery from unusual attitudes.

Completion Standards. The lesson will have been successfully completed when, by an oral test, the student displays a basic knowledge of the performance of takeoffs and landings under various conditions, and an understanding of the performance of basic maneuvers by reference to flight instruments.

Lesson No. 16

Objective. To develop the student's knowledge of the fundamentals of night flying.

Content.

1. Night flying—general.
   a. Requirements of regulations.
   b. Preparation.
   c. Equipment.
   d. Night vision.
   e. Airport lighting.
   f. Orientation.
   g. VFR navigation.
   h. Weather factors.
2. Partial or complete power failure.
   a. Sample situations.
   b. Recommended courses of action.
3. Systems and equipment malfunctions.

a. Sample situations.

b. Recommended courses of action.

Completion Standards. The lesson will have been successfully completed when, by an oral test, the student displays a working knowledge of the fundamentals of night flying.

PART 2: A FLIGHT INSTRUCTION CURRICULUM

The sample private pilot flight training lessons which appear on the following pages are illustrative of course content and organization. It is not necessarily the most desirable syllabus to use in all training situations; however, instruction in the procedures and maneuvers listed here are considered to be most effective in the development of competence in the pilot operations required on the private pilot (airplane) flight test.

It should be noted that each lesson prescribes a unit of flight training, not a specified period of instruction or flight time. The notation "(VR and IR)" is used to indicate maneuvers which should be performed by both visual references and instrument references during the conduct of flight instruction.

Lesson No. 1—Dual

Objective. To familiarize the student with the training airplane, its servicing, its operating characteristics, cabin controls, instruments, systems, preflight procedures, use of checklists, and safety precautions to be followed; to acquaint the student with the sensations of flight and the effect and use of controls; and to familiarize the student with the local flying area and airport.

Content.

1. Preflight discussion.
2. Introduction.
   a. Airplane servicing.
   b. Purpose of preflight checks.
   c. Visual inspection.
   d. Importance of using a checklist.
   e. Engine starting procedure.
   f. Radio communications procedures.
   g. Taxiing.
   h. Pretakeoff checklist.
   i. Takeoff.
   j. Traffic pattern departure, climb-out, and level-off.
   k. Effect and use of controls.

l. Straight-and-level flight.
m. Medium bank turns.
n. Local flying area familiarization.
o. Collision avoidance.
p. Wake turbulence avoidance.
q. Traffic pattern entry, approach, landing, and parking.
r. Ground safety.
3. Postflight critique and preview of next lesson.

Completion Standards. The lesson will have been successfully completed when the student understands how to service the airplane, the use of a checklist for the visual inspection, starting procedure, and engine runup; displays a knowledge of the effect and use of controls; and has a reasonable familiarity with the local flying area and airport.

Lesson No. 2—Dual

Objective. To develop the student's skill in the performance of the four basic flight maneuvers (climbs, descents, turns, and straight and level flight).

Content.
1. Preflight discussion.
2. Review.
   a. Airplane servicing.
   b. Visual inspection.
   c. Engine starting procedure.
   d. Radio communications procedures.
   e. Taxiing.
   f. Pretakeoff checklist.
   g. Takeoff.
   h. Traffic pattern departure.
   i. Straight-and-level flight (VR and IR).
   j. Medium bank turns (VR and IR).
   k. Traffic pattern entry, approach, landing, and parking.
3. Introduction.
   a. Climbs and climbing turns.
   b. Glides and gliding turns.
   c. Torque effect.
   d. Level-off from climbs and glides.
4. Postflight critique and preview of next lesson.

Completion Standards. The lesson will have been successfully completed when the student can perform, with minimum assistance from the instructor, climbs, straight-and-level flight, turns, and glides. During straight-and-level flight the student should, with minimum instructor assistance, be able to maintain altitude within ±100 feet, airspeed within ±10 knots, and heading within ±10° of that assigned.

Lesson No. 3—Dual

Objective. To review lessons One and Two; to develop the student's proficiency in the performance of the basic flight maneuvers; and to introduce maneuvering at minimum controllable airspeed and power-off stalls.

Content.
1. Preflight discussion.
2. Review.
   a. Use of checklist.
   b. Engine starting procedure.
   c. Radio communications procedures.
   d. Takeoff.
   e. Traffic pattern departure.
   f. Climbs and climbing turns (VR and IR).
   g. Straight-and-level flight (VR and IR).
   h. Medium bank turns (VR and IR).
   i. Glides and gliding turns (VR and IR).
   j. Level-off procedures (VR and IR).
   k. Traffic pattern and landing.
3. Introduction.
   a. Maneuvering at minimum controllable airspeed.
   b. Power-off stalls (imminent and full).
   c. Descents and descending turns, with power.
4. Postflight critique and preview of next lesson.

Completion Standards. The lesson will have been successfully completed when the student can display reasonable proficiency in the performance of the four basic flight maneuvers, and perform with minimum assistance, flight at minimum controllable airspeed. During this and subsequent flight lessons, the student should be able to perform the visual inspection, starting procedure, radio communications, taxiing, pretakeoff check, parking, and shut-down procedure without assistance. During climbs, level flight, turns, glides, and maneuvering at minimum controllable airspeed the student should, with minimum instructor assistance, be able to maintain assigned airspeed within ±10 knots. The student should also, with minimum instructor assistance, be able to maintain assigned altitude within ±100 feet and assigned heading within ±10°.

Lesson No. 4—Dual

Objective. To review previous lessons, thereby increasing the student's competence in the performance of fundamental flight maneuvers; and to introduce power-on stalls, rectangular course, S-turns across a road, eights along a road, and elementary emergency landing.

Content.
1. Preflight discussion.
2. Review.
   a. Takeoff.
   b. Traffic pattern departure.
   c. Climbs and climbing turns (VR and IR).
   d. Straight-and-level flight and medium bank turns (VR and IR).
   e. Maneuvering at minimum controllable airspeed (VR and IR).
   f. Power-off stalls (imminent and full) (VR and IR).
   g. Glides and gliding turns (VR and IR).
   h. Descents and descending turns, with power (VR and IR).
   i. Level-off procedures (VR and IR).
   j. Traffic pattern and landing.
3. Introduction.
   a. Power-on stalls (imminent and full).
   b. Rectangular course.
   c. S-turns across a road.
   d. Eights along a road and eights across a road.
   e. Elementary emergency landings.
4. Postflight critique and preview of next lesson.

Completion Standards. The lesson will have been successfully completed when the student is competent to perform, with minimum instructor assistance, the procedures and maneuvers given during previous lessons. The student should achieve the ability to recognize stall indications and make safe prompt recoveries. The student should maintain assigned airspeed within ±10 knots, assigned altitude within ±100 feet, and assigned heading within ±10°, and display a basic knowledge of elementary emergency landings.

Lesson No. 5—Dual

Objective. To review previous lessons, with emphasis on maneuvering by reference to ground objects. To develop the student's ability to perform climbs at best rate and best angle, crosswind takeoffs and landings; and to introduce emergency procedures, changes of airspeed and configuration, turns around a point, and eights around pylons.

Content.

1. Preflight discussion.
2. Review.
   a. Takeoff.
   b. Climbs and climbing turns (VR and IR).
   c. Maneuvering at minimum controllable airspeed (VR and IR).
   d. Power-off and power-on stalls (imminent and full).
   e. Rectangular course.
   f. S-turns across a road.
   g. Eights along a road.
   h. Elementary emergency landings.
   i. Traffic pattern and landing.
3. Introduction.
   a. Crosswind takeoffs and landings.
   b. Climb at best rate.
   c. Climb at best angle.
   d. Emergency procedures.
   e. Change of airspeed and configuration (VR and IR).
   f. Turns around a point.
   g. Eights around pylons.
4. Postflight critique and preview of next lesson.

Completion Standards. The lesson will have been successfully completed when the student can recognize imminent and full stalls and make prompt effective recoveries, perform ground reference maneuvers with reasonably accurate wind drift corrections and good coordination, and has a proper concept of crosswind technique during takeoffs and landings. The student should have a working knowledge of emergency procedures, and be able to perform them with minimum assistance. During ground reference maneuvers, the student should maintain airspeed within ±10 knots, altitude within ±100 feet, and heading within ±10° of that desired.

Lesson No. 6—Dual

Objective. To review previous lessons; to develop the student's ability to perform slips, accelerated stalls, cross-control stalls, and advanced emergency landings; to improve the student's proficiency in normal and crosswind takeoffs and landings; and to introduce balked takeoffs and go-arounds (rejected landings).

Content.

1. Preflight discussion.
2. Review.
   a. Normal and crosswind takeoffs.
   b. Climbs at best rate and best angle.
   c. Power-off stalls (imminent and full) (VR and IR).
   d. Power-on stalls (imminent and full) (VR and IR).
   e. Change of airspeed and configuration (VR and IR).
   f. Turns around a point.
   g. Eights around pylons.
   h. Emergency procedures.
   i. Normal and crosswind landings.
3. Introduction.
   a. Balked takeoffs.
   b. Accelerated stalls.
   c. Cross-control stalls.
   d. 180° and 360° gliding approaches.
   e. Advanced emergency landings.
   f. Side slips and forward slips.
   g. Go-arounds (rejected landings).
4. Postflight critique and preview of next lesson.

Completion Standards. The lesson will have been successfully completed when the student can perform stall recoveries smoothly and promptly with a minimum loss of altitude, is able to make unassisted normal and crosswind takeoffs and landings, and can plan and fly emergency landing patterns with accuracy and consistency. The student should be able to execute balked takeoffs and go-arounds (rejected landings) without assistance, and should maintain assigned airspeed within ±10 knots, assigned altitude within ±100 feet, and assigned heading within ±10°.

Lesson No. 7—Dual

Objective. To review previous lessons. To further develop the student's competence in takeoffs, traffic patterns, and landings through concentrated practice. To develop the student's ability to use slips during landing approaches and improve the ability to perform go-arounds (rejected landings).

Content.

1. Preflight discussion.

2. Review.
   a. Normal and crosswind takeoffs.
   b. Normal and crosswind landings (touch-and-go and full-stop).
   c. Forward slips.
   d. Go-arounds (rejected landings).
   e. 180° and 360° gliding approaches.
   f. Advanced emergency landings.
   g. Emergency procedures.
3. Postflight critique and preview of next lesson.

Completion Standards. The lesson will have been successfully completed when the student can fly accurate traffic patterns and make unassisted normal and crosswind takeoffs and landings. The student should be competent in the go-around (rejected landing) procedure. During traffic patterns, the student should maintain desired airspeed within ±10 knots, desired altitude within ±100 feet, and desired heading within ±10°.

Lesson No. 8—Dual

Objective. To review power-off stalls, maneuvering at minimum controllable airspeed, and advanced emergency landings. To continue to develop the student's competence in takeoffs, traffic patterns, and landings, and to improve the ability to recover from poor approaches and landings.

Content.
1. Preflight discussion.
2. Review.
   a. Normal and crosswind takeoffs.
   b. Power-off stalls (imminent and full) (VR and IR).
   c. Maneuvering at minimum controllable airspeed (VR and IR).
   d. Advanced emergency landings.
   e. Normal and crosswind landings (touch-and-go and full-stop).
   f. Go-arounds (rejected landings).
   g. Recovery from poor approaches and landings.
3. Post flight critique and preview of next lesson.

Completion Standards. The lesson will have been successfully completed when the student can demonstrate a degree of proficiency in normal and crosswind takeoffs and landings and traffic patterns, which is considered safe for solo. The student should display sound judgment and proper techniques in recoveries from poor approaches and landings. During traffic patterns, the student should maintain desired airspeed within ±10 knots, desired altitude within ±100 feet, and desired heading within ±10°.

Lesson No. 9 — Dual and Solo

Objective. To develop the student's competence to a level which will allow the safe accomplishment of the first supervised solo in the traffic pattern.

Content.
1. Preflight discussion.
2. Review.
   a. Normal and crosswind takeoffs.
   b. Normal and crosswind landings (full stop).
   c. Go-arounds (rejected landings).
   d. Recovery from poor approaches and landings.
   e. Elementary emergency landings.
3. Introduction—first supervised solo in the traffic pattern. Three takeoffs and three full-stop landings should be performed.
4. Postflight critique and preview of next lesson.

Completion Standards. The lesson will have been successfully completed when the student safely accomplishes the first supervised solo in the traffic pattern.

Lesson No. 10 — Dual and Solo

Objective. To review previous lessons and to accomplish the student's second supervised solo in the traffic pattern.

Content.
1. Preflight discussion.
2. Review.
   a. Takeoff and traffic departure.
   b. Climbs and climbing turns (VR and IR).
   c. Maneuvering at minimum controllable airspeed (VR and IR).
   d. Power-off stalls (imminent and full) (VR and IR).
   e. Advanced emergency landings.
   f. Traffic patterns, approaches and landings.
   g. Recovery from poor approaches and landings.
3. Introduction—second supervised solo in the traffic pattern. Three takeoffs, two touch-and-go, and one full-stop landing should be performed.
4. Postflight critique and preview of next lesson.

Completion Standards. The lesson will have been successfully completed when the student demonstrates solo competence in maneuvers performed and safely accomplishes the second supervised solo in the traffic pattern.

Lesson No. 11—Dual and Solo

Objective. To review presolo maneuvers with higher levels of proficiency required. To introduce short and soft field takeoffs, and maximum climbs; and to accomplish the student's third supervised solo in the traffic pattern.

Content.
1. Preflight discussion.
2. Review.
   a. Selected presolo maneuvers (VR and IR).
   b. Takeoffs, traffic patterns, and landings.
   c. Balked takeoff.
   d. Go-around (rejected landing).
   e. Recovery from poor approach and landing.
3. Introduction.
   a. Short field takeoffs and maximum climbs.
   b. Soft field takeoffs.
   c. Third supervised solo in the traffic pattern. At least three takeoffs and landings should be performed.
4. Preflight critique and preview of next lesson.

Completion Standards. The lesson will have been successfully completed when the student demonstrates solo competence in the selected presolo maneuvers performed and safely accomplishes the third supervised solo in the traffic pattern. The student should be able to perform short field takeoffs, soft field takeoffs, and maximum climbs without instructor assistance.

Lesson No. 12—Dual

Objective. To refamiliarize the student with the local practice area and to improve proficiency in the presolo maneuvers in preparation for local area solo practice flights. To develop the student's ability to obtain radar and DF heading instructions and to become oriented in relation to a VOR, and to "home" to a nondirectional beacon using ADF. To introduce wheel landings (tail wheel airplanes).

Content.
1. Preflight discussion.
2. Review.
   a. Practice area orientation.
   b. Power-off stalls (imminent and full) (VR and IR).
   c. Power-on stalls (imminent and full) (VR and IR).
   d. Maneuvering at minimum controllable airspeed (VR and IR).
   e. Turns around a point.

f. Eights around pylons.

g. Crosswind takeoffs and landings.

h. 180° and 360° gliding approaches.

i. Advanced emergency landings.

j. Emergency procedures.

3. Introduction.

a. Use of radar and DF heading instructions.

b. VOR orientation.

c. ADF "homing."

d. Wheel landings (tailwheel airplanes).

4. Postflight critique and preview of next lesson.

Completion Standards. The lesson will have been successfully completed when the student demonstrates an improved performance of the presolo maneuvers, is able to determine position in the local practice area by pilotage, VOR, or ADF; and can safely perform signed maneuvers. The student should be competent in obtaining radar and DF heading instructions and in the performance of simulated emergency landings and emergency procedures.

Lesson No. 13—Solo

Objective. To develop the student's confidence and proficiency through solo practice of assigned maneuvers.

Content.

1. Preflight discussion.

2. Review.

a. Normal and/or crosswind takeoffs and landings.

b. Lower-off stalls (imminent and full).

c. Power-on stalls (imminent and full).

d. Maneuvering at minimum controllable airspeed.

e. Other maneuvers specified by the instructor during the preflight discussion.

3. Postflight critique and preview of next lesson.

Completion Standards. The lesson will have been successfully completed when the student has accomplished the solo review and practiced the basic and precision flight manuevers, in addition to those specified by the instructor. The student should gain confidence and improve flying technique as a result of the solo practice period.

Lesson No. 14—Dual

Objective. To improve the student's proficiency in previously covered procedures and maneuvers and to review advanced emergency landings, emergency procedures, and orientation by means of VOR and/or ADF.

Content.
1. Preflight discussion.
2. Review.
   a. Normal and/or crosswind takeoffs and landings.
   b. Power-off stalls (imminent and full) (VR and IR).
   c. Power-on stalls (imminent and full) (VR and IR).
   d. Maneuvering at minimum controllable airspeed (VR and IR).
   e. Accelerated stalls.
   f. Eights around pylons.
   g. Short field and soft field takeoffs and landings.
   h. Advanced emergency landings.
   i. Emergency procedures.
   j. Orientation by means of VOR and/or ADF.
3. Postflight critique and preview of next lesson.

Completion Standards. The lesson will have been successfully completed when the student demonstrates an increased proficiency in previously covered procedures and maneuvers. The student should be able to maintain airspeed within ±10 knots, altitude within ±100 feet, and heading within ±10° of that desired.

Lesson No. 15—Solo

Objective. To further develop the student's confidence and proficiency through solo practice of assigned maneuvers.

Content.
1. Preflight Discussion.
2. Review.
   a. Normal and/or crosswind takeoffs and landings.
   b. Turns around a point.
   c. Eights around pylons.
   d. Other maneuvers specified by the instructor during the preflight discussion.
3. Postflight critique and preview of next lesson.

Completion Standards. The lesson will have been successfully completed when the student has accomplished the solo review and thereby increased proficiency and confidence.

Lesson No. 16—Dual

Objective. To develop the student's ability to plan, plot, and fly a

144

2-hour day cross-country flight with landings at two unfamiliar airports; to develop the student's proficiency in navigating by means of pilotage, dead reckoning, VOR, and/or ADF; and to develop the ability to take proper action in emergency situations.

Content.

1. Preflight discussion.
    a. Planning flight, including weather check.
    b. Plotting course.
    c. Preparing log.
    d. Filing and closing VFR flight plan.
2. Introduction.
    a. Filing VFR flight plan.
    b. Pilotage.
    c. Dead reckoning.
    d. Tracking VOR radial and/or homing by ADF.
    e. Departure, en route, and arrival radio communications.
    f. Simulated diversion to an alternate airport.
    g. Unfamiliar airport procedures.
    h. Emergencies, including DF and radar heading instructions.
    i. Closing VFR flight plan.
3. Postflight critique and preview of next lesson.

Completion Standards. The lesson will have been successfully completed when, with instructor assistance, the student is able to perform the cross-country preflight planning, fly the planned course making necessary off-course corrections, and can make appropriate radio communications. The student should be competent in navigating by means of pilotage, dead reckoning, VOR, and/or ADF, and when so instructed, is able to accurately plan and fly a diversion to an alternate airport.

Lesson No. 17—Dual

Objective. To improve the student's proficiency in cross-country operations through the planning, plotting, and flying of a second dual 2-hour day cross-country flight, with landings at two unfamiliar airports. To improve the student's competence in navigating by means of pilotage, dead reckoning, VOR, and ADF; and to further develop the ability to take proper action in emergency situations.

Content.

1. Preflight discussion.
    a. Planning flight, including weather check.
    b. Plotting course.

c. Preparing log.

d. Filing and closing VFR flight plan.

2. Review.

   a. Filing VFR flight plan.

   b. Pilotage and dead reckoning.

   c. Radio navigation (VOR and/or ADF).

   d. Departure, en route, and arrival radio communications.

   e. Simulated diversion to an alternate airport.

   f. Unfamiliar airport procedures.

   g. Emergencies, including DF and radar heading instructions (VR and IR).

   h. Closing VFR flight plan.

3. Postflight critique and preview of next lesson.

Completion Standards. The lesson will have been successfully completed when the student, with minimum instructor assistance, is able to plan, plot, and fly the planned course. Estimated times of arrival should be accurate with an apparent error of not more than 10 minutes. Any off-course corrections should be accomplished accurately and promptly. The student should be able to give the instructor an accurate position report at any time without hesitation. When given a "simulated lost" situation, the student should be able to initiate and follow an appropriate "lost procedure."

Lesson No. 18—Solo

Objective. To develop the student's ability to plan, plot, and fly a 3-hour solo day cross-country flight, with landings at two unfamiliar airports, thereby improving proficiency and confidence in the conduct of future solo cross-country flights. To improve the student's proficiency in navigating by means of pilotage, dead reckoning, VOR, and/or ADF; and to increase the ability to cope with new or unexpected flight situations.

Content.

1. Preflight discussion.

   a. Planning flight, including weather check.

   b. Plotting course.

   c. Preparing log.

   d. Filing and closing VFR flight plan.

   e. Procedure at unfamiliar airports.

   f. Emergencies.

2. Review.

   a. Filing VFR flight plan.

   b. Pilotage.

c. Dead reckoning.

d. Radio navigation (VOR and/or ADF).

e. Departure, en route, and arrival radio communications.

f. Unfamiliar airport procedures.

g. Closing VFR flight plan.

3. Postflight critique and preview of next lesson.

Completion Standards. The lesson will have been successfully completed when the student is able to plan, plot, and fly the 3-hour cross-country flight as assigned by the instructor. The instructor should determine how well the flight was conducted through oral questioning.

Lesson No. 19—Dual and Solo

Objective. To develop the student's ability to make solo night flights in the local practice area and airport traffic pattern. To familiarize the student with such aspects of night operations as: night vision, night orientation, judgment of distance, use of cockpit lights, position lights, landing lights, and night emergency procedures.

Content.

1. Preflight discussion.

    a. Night vision and vertigo.

    b. Orientation in local area.

    c. Judgment of distance.

    d. Aircraft lights.

    e. Airport lights.

    f. Taxi technique.

    g. Takeoff and landing technique.

    h. Collision avoidance.

    i. Unusual attitude recovery.

    j. Emergencies.

2. Introduction.

    a. Night visual inspection.

    b. Use of cockpit lights.

    c. Taxi techniques.

    d. Takeoff and traffic departure.

    e. Area orientation.

    f. Interpretation of aircraft and airport lights.

    g. Recovery from unusual attitudes.

    h. Radio communications.

    i. Traffic entry.

    j. Power approaches and full-stop landings.

k. Use of landing lights.

l. Simulated electrical failure to include at least one black-out landing.

3. Postflight critique and preview of next lesson.

Completion Standards. The lesson will have been successfully completed when the student displays the ability to maintain orientation in the local flying area and traffic pattern, can accurately interpret aircraft and runway lights, and can competently fly the traffic pattern and perform takeoffs and landings. The student should display, through oral quizzing and demonstrations, competence in performing night emergency procedures. At least five takeoffs and landings should be accomplished.

Lesson No. 20—Dual

Objective. To develop the student's ability to plan, plot, and fly a 1½-hour night cross-country flight around a triangular course with at least one landing at an unfamiliar airport. To develop the student's competence in navigating at night by means of pilotage, dead reckoning, and VOR or ADF; and to develop the student's ability to take proper action in night emergency situations.

Content.

1. Preflight discussion.

   a. Planning 1½-hour night cross-country flight, including weather check.

   b. Plotting course.

   c. Preparing log.

   d. Filing and closing VFR flight plan.

2. Introduction.

   a. Filing VFR flight plan.

   b. Proper use of cockpit lights and flashlight for chart reading.

   c. Pilotage—factors peculiar to night flying.

   d. Dead reckoning.

   e. Tracking VOR radial and/or homing by ADF.

   f. Departure, en route, and arrival radio communications.

   g. Simulated diversion to an alternate airport.

   h. Emergencies, including simulated failure of electrical system, also DF and radar heading instructions.

   i. Closing VFR flight plan.

3. Postflight critique and preview of next lesson.

Completion Standards. The lesson will have been successfully completed when, with minimum assistance from the instructor, the student is able to perform the night cross-country preflight planning, fly the planned course making necessary off-course corrections, and can make appropriate radio communications. The student should be competent in navigating by means of pilotage, dead reckoning, and VOR or ADF. The student should have a thorough knowledge of night emergency procedures.

Lesson No. 21—Solo

Objective. To further develop the student's competence in cross-country operations through the planning, plotting, and flying of a second solo 3-hour day cross-country flight with landings at two unfamiliar airports. To improve the student's proficiency in navigating by means of pilotage, dead reckoning, VOR, and/or ADF; and to further increase the student's confidence and ability to properly handle unexpected flight situations.

Content.
1. Preflight discussion.
   a. Planning flight, including weather check.
   b. Plotting course.
   c. Preparing log.
   d. Filing and closing VFR flight plan.
   e. Procedure at unfamiliar airports.
   f. Emergencies.
2. Review.
   a. Filing VFR flight plan.
   b. Pilotage and dead reckoning.
   c. Radio navigation (VOR and/or ADF).
   d. Departure, en route, and arrival radio procedures.
   e. Unfamiliar, airport procedures.
   f. Closing VFR flight plan.
3. Postflight critique and preview of next lesson.

Completion Standards. The lesson will have been successfully completed when the student is able to plan, plot, and fly the second 3-hour day cross-country flight as assigned by the instructor. The instructor should determine how well the flight was conducted through oral questioning.

Lesson No. 22—Solo

Objective. To further develop the student's competence in cross-country operations through the planning, plotting, and flying of a solo 4-hour day cross-country flight, with landings at three unfamiliar airports, each of which is more than 100 nautical miles from the other airports.

Content.
1. Preflight discussion.
   a. Planning flight, including weather check.
   b. Plotting course.
   c. Preparing log.
   d. Filing and closing VFR flight plan.
   e. Procedure at unfamiliar airports.
   f. Emergencies.
2. Review.
   a. Filing VFR flight plan.
   b. Pilotage and dead reckoning.
   c. Radio navigation (VOR and/or ADF).
   d. Departure, en route, and arrival radio communications.
   e. Unfamiliar, airport procedures.
   f. Closing VFR flight plan.
3. Postflight critique and preview of next lesson.

Completion Standards. The lesson will have been successfully completed when the student is able to plan, plot, and fly the second 4-hour day cross-country flight as assigned by the instructor. The instructor should determine how well the flight was conducted through oral questioning.

Lesson No. 23—Dual

Objective. To develop precision in the student's performance of procedures and maneuvers covered previously with emphasis directed to stalls.

Content.
1. Preflight discussion.
2. Review.
   a. Power-off stalls (imminent and full) (VR and IR).
   b. Power-on stalls (imminent and full) (VR and IR).
   c. Maneuvering at minimum controllable airspeed (VR and IR).
   d. 180° and 360° gliding approaches.
   e. Advanced emergency landings.
   f. Slips.
   g. Crosswind takeoffs and landings.
   h. Short field and soft field takeoffs and landings.
   i. Emergency procedures.
3. Introduction of ASR approaches.
4. Postflight critique and preview of next lesson.

Completion Standards. The lesson will have been successfully completed when the student demonstrates improved performance in the various maneuvers given. The student should be able to make ASR approaches with minimum instructor assistance.

Lesson No. 24—Solo

Objective. To further develop the student's competence through solo practice of assigned maneuvers. Emphasis will be directed to stalls.

Content.
1. Preflight discussion.
2. Review.
   a. Power-on and power-off stalls (imminent and full).
   b. Maneuvering at minimum controllable airspeed.
   c. Short field and soft field takeoffs and landings.
   d. Other maneuvers assigned by.the instructor during preflight discussion.
3. Postflight critique and preview of next lesson.

Completion Standards. The lesson will have been successfully completed when the student has accomplished the solo review and practiced the basic and precision flight maneuvers in addition to those specified by the instructor. The student should gain confidence and improve flying technique as a result of the solo practice period.

Lesson No. 25—Dual

Objective. To develop improved performance and precision in the procedures and maneuvers covered previously with emphasis directed to ground track maneuvers.

Content.
1. Preflight discussion.
2. Review.
   a. Maneuvering at minimum controllable airspeed.
   b. Turns around a point.
   c. Eights around pylons.
   d. 180° and 360° gliding approaches.
   e. Advanced emergency landings.
   f. Slips.
   g. Crosswind takeoffs and landings.
   h. Wheel landings (tail wheel airplane).
   i. ASR approach (facilities permitting).
3. Postflight critique and preview of next lesson.

Completion Standards. The lesson will have been successfully completed when the student demonstrates improved performance in the maneuvers given.

Lesson No. 26—Solo

Objective. To further develop the student's competence through solo practice of assigned maneuvers. Emphasis will be directed to ground track maneuvers.

Content.
1. Preflight discussion.
2. Review.
   a. Turns around a point.
   b. Eights around plyons.
   c. Short and soft field takeoffs and landings.
   d. Wheel landings (tail wheel airplanes).
   e. Other maneuvers assigned by the instructor during the preflight discussion.
3. Postflight critique and preview of next lesson.

Completion Standards. The lesson will have been successfully completed when the student has accomplished the solo review. The student should gain proficiency in the ground track and other maneuvers assigned by the instructor.

Lesson No. 27—Solo

Objective. To improve the student's proficiency in the pilot operations required on the private pilot (airplane) flight check.

Content.
1. Preflight discussion.
2. Review.
   a. Ground track maneuvers.
   b. Power-on and power-off stalls (imminent and full).
   c. Maneuvering at minimum controllable airspeed.
   d. Crosswind takeoffs and landings.
   e. Other maneuvers assigned by the instructor during the preflight discussion.
3. Post flight critique and preview of next lesson.

Completion Standards. The lesson will have been successfully completed when the student has gained proficiency in the procedures and maneuvers assigned by the instructor.

Lesson No. 28—Dual

Objective. To evaluate the student's performance of the procedures and maneuvers necessary to conduct flight operations as a private pilot.

Content.

1. Preflight discussion.
2. Review.
   a. Power-on and power-off stalls (imminent and full).
   b. Maneuvering at minimum controllable airspeed.
   c. Ground track maneuvers.
   d. 180° and 360° gliding approaches.
   e. Advanced emergency landings.
   f. Short field and soft field takeoffs and landings.
   g. Crosswind takeoffs and landings.
   h. Straight-and-level flight, turns, climbs, descents, and recovery from unusual attitudes by reference to flight instruments.
   i. Tracking VOR radial and homing by ADF (VR and IR).
   j. Use of radar and DF heading instructions (VR and IR).
   k. ASR approach (VR and IR).
   l. Emergency operations.

Completion Standards. The lesson will have been successfully completed when the student satisfactorily performs the procedures and maneuvers selected to show competence in the pilot operations listed in the Private Pilot (Airplane) Flight Test Guide.

PART 3: THE PRIVATE PILOT FLIGHT TEST

(Extracted from the *Private Pilot-Airplane Flight Test Guide*, AC 61-54A.)

Use of the Flight Test Guide

The pilot operations in this flight test guide, indicated by Roman numerals, are required by Section 61.107 of Part 61 (revised) for the issuance of a Private Pilot Certificate. For the addition of a class rating on the pilot certificate, an applicant must pass a flight test appropriate to the pilot certificate and applicable to the aircraft category and class rating sought. This guide is intended only to outline appropriate pilot operations and the minimum standards for the performance of each procedure or maneuver which will be accepted by the examiner as evidence of the pilot's competency. It is not intended that the applicant be tested on every procedure or maneuver within each pilot operation, but only those considered necessary by the examiner to determine competency in each pilot operation.

When, in the judgment, of the examiner, certain demonstrations are impractical (for example, night flying or equipment malfunctions), competency may be determined by oral testing.

This guide contains an *objective* for each required pilot operation. Under each pilot operation, pertinent procedures or maneuvers are listed with *Descriptions* and *Acceptable Performance Guidelines.*

1. The *Objective* states briefly the purpose of each pilot operation required on the flight test.

2. The *Description* provides information on what may be asked of the applicant regarding the selected procedure or maneuver. The procedures or maneuvers listed have been found most effective in demonstrating the objective of that particular pilot operation.

3. The *Acceptable Performance Guidelines* include the factors which will be taken into account by the examiner in deciding whether the applicant has met the objective of the pilot operation. The airspeed, altitude, and heading tolerances given represent the minimum performance expected in good flying conditions. However, consistently exceeding these tolerances before corrective action is initiated is indicative of an unsatisfactory performance. Any procedure or action, or the lack thereof, which requires the intervention of the examiner to maintain safe flight will be disqualifying. Failure to exercise proper vigilance or to take positive action to ensure that the flight area has been adequately cleared for conflicting traffic will also be disqualifying.

Emphasis will be placed on procedures, knowledge, and maneuvers which are most critical to a safe performance as a pilot. The demonstration of prompt stall recognition, adequate control, and recovery techniques will receive special attention. Other areas of importance include spatial disorientation, collision avoidance, and wake turbulence hazards.

The applicant will be expected to know the meaning and significance of the airplane performance speeds important to the pilot, and be able to readily determine these speeds for the airplane used for the flight test. These speeds include:

V_{so} —the stalling speed or minimum steady flight speed in landing configuration.

V_y —the speed for the best rate of climb.

V_x —the speed for the best angle of climb.

V_a —the design maneuvering speed.

V_{ne} —the never exceed speed.

General Procedures for Flight Tests

The ability of an applicant for a private or commercial pilot certificate, or for an aircraft or instrument rating on that certificate, to perform the required pilot operations is based on the following:

1. Executing procedures and maneuvers within the aircraft's performance capabilities and limitations, including use of the aircraft's systems.

2. Executing emergency procedures and maneuvers appropriate to the aircraft.

3. Piloting the aircraft with smoothness and accuracy.

4. Exercising judgment.

5. Applying aeronautical knowledge.

6. Showing mastery of the aircraft, with the successful outcome of a procedure or maneuver never seriously in doubt.

If the applicant fails any of the required pilot operations, the flight test is failed. The examiner or the applicant may discontinue the test at any time when the failure of a required pilot operation makes the applicant ineligible for the certificate or rating sought. If the test is discontinued the applicant is entitled to credit for only those entire pilot operations that were successfully performed.

Flight Test Prerequisites

An applicant for the private pilot flight test is required by revised Section 61.39 of the Federal Aviation Regulations to have: (1) passed the appropriate private pilot written test within 24 months before the date of the flight test, (2) the applicable instruction and aeronautical experience prescribed for a private pilot certificate, (3) a first, second, or third class medical certificate issued within the past 24 months, (4) reached at least 17 years of age, and (5) a written statement from an appropriately certificated flight instructor certifying that he has given the applicant flight instruction in preparation for the flight test within 60 days preceding the date of application, and finds that person competent to pass the test and to have a satisfactory knowledge of the subject areas in which a deficiency is shown by the airman written test report.

Airplane and Equipment Requirements for Flight Test

The applicant is required by revised Section 61.45 to provide an airworthy airplane for the flight test. This airplane must be capable of, and its operating limitations must not prohibit, the pilot operations required in the test. The following equipment is relevant to the pilot operations required by revised Section 61.107 for the private pilot flight test:

1. Two-way radio suitable for voice communications with aeronautical ground stations.

2. A radio receiver which can be utilized for available radio navigation and communications facilities.

3. Appropriate flight instruments for controlling and maneuvering an airplane solely by reference to instruments.

The inspector/examiner may accept any aircraft that, in his judgment, is adequately equipped to evaluate the applicant's instrument training to the standards outlined in this document.

4. Engine and flight controls that are easily reached and operated in a normal manner by both pilots.

5. A suitable view-limiting device, easy to install and remove in flight, for simulating instrument flight conditions.

6. Operating instructions and limitations. The applicant should have an appropriate checklist, an Owner's Manual/ Handbook, or, if required for the airplane used, an FAA approved Airplane Flight Manual. Any operating limitations or other published recommendations of the manufacturer that are applicable to the specific airplane will be observed.

I. Preflight Operations

Objective: To determine that the applicant can ensure meeting the pilot requirements, that the airplane is airworthy and ready for safe flight, and that suitable weather conditions exist for the proposed flight.

A. Certificates and Documents.

1. Description. The applicant may be asked to present appropriate pilot and medical certificates and to locate and explain the airplane's registration certificate, airworthiness certificate, operating manual or FAA approved Airplane Flight Manual (if required), airplane equipment list, and required weight and balance data. In addition the applicant must be able to explain the airplane and engine logbooks or other maintenance records.

2. Acceptable Performance Guidelines. The applicant shall be knowledgeable regarding the location, purpose, and significance of each required item.

156

B. Airplane Performance and Limitations.

1. Description. The applicant may be orally quizzed on the performance capabilities, approved operating procedures, and limitations of the airplane used. This includes normal power settings, critical and recommended speeds, and fuel and oil requirements. In addition, the manufacturer's published recommendations[2] or FAA approved Airplane Flight Manual should be used to determine the effects of temperature, pressure altitude, wind, and gross weight on the airplane's performance.

2. Acceptable Performance Guidelines. The applicant shall be evaluated on the ability to obtain, explain, and apply the information which is essential in determining the performance capabilities and limitations of the airplane used.

C. Weight and Balance.

1. Description. The applicant may be asked to demonstrate the application of the approved weight and balance data for the airplane used to determine that the gross weight and c.g. (center of gravity) location are within allowable limits. Charts and graphs provided by the manufacturer may be used.

2. Acceptable Performance Guidelines. The applicant's performance shall be evaluated on ability to determine the empty weight, c.g., maximum allowable gross weight, useful load (fuel, passengers, baggage) by reference to appropriate publications, and the ability to apply this information to determine that the gross weight and c.g. are within approved limits.

D. Weather Information.

1. Description. The applicant may be asked to obtain Aviation Weather Reports, Area and Terminal Forecasts, and Winds Aloft Forecasts pertinent to the proposed flight.

2. Acceptable Performance Guidelines. The applicant shall demonstrate knowledge to select weather information that is pertinent and how to best obtain this information. Applicant must show ability to interpret and understand its significance with respect to the proposed flight.

E. Line Inspection.

1. Description. The applicant may be asked to demonstrate a visual inspection to determine the airplane's airworthi-

---

[2]The phrase "manufacturer's published recommendations" is used hereafter in this guide to denote FAA approved Airplane Flight Manual material when such material has been approved for the airplane type or other manufacturer's published recommendations such as "Owner's Manual," "Owner's Handbook," "Bulletins," and "Letters" for the safe operation of the airplane model or series, in the absence of an approved Airplane Flight Manual.

ness and readiness for flight. This includes all required equipment and documents. A checklist provided by the manufacturer or operator should be used.

2. Acceptable Performance Guidelines. The applicant shall use an orderly procedure in conducting a preflight check of the airplane, and shall know the significance of each item checked and recognize any unsafe condition.

F. Airplane Servicing.

1. Description. The applicant may be asked to demonstrate a visual inspection to determine that the fuel is of the proper grade and type and the supply of fuel, oil, and other required fluids is adequate for the proposed flight. Appropriate action should be taken to eliminate possible fuel contamination in the airplane.

2. Acceptable Performance Guidelines. The applicant shall know the grade and type of oil and fuel specified for the airplane and be able to determine the amount of fuel required to complete the flight. Applicant shall know where to find all fuels and oil filters, and the capacity of each tank, as well as the location of the battery, hydraulic fluid reservoirs, anti-icing fluid tanks, etc., and shall also know the proper steps for avoiding fuel contamination during the following servicing.

G. Engine and Systems Preflight Check.

1. Description. The applicant may be asked to demonstrate a check to determine that the engine is operating within acceptable limits and that all systems, equipment, and controls are functioning properly and adjusted for takeoff. A checklist provided by the manufacturer or operator should be used.

2. Acceptable Performance Guidelines. The applicant shall use proper procedures in engine starting and runup and in checking airplane systems, equipment, and controls to determine that the airplane is ready for flight. Careless operation in close proximity to obstructions, ground personnel, or other aircraft shall be disqualifying.

II. Airport and Traffic Pattern Operations

Objective: To determine that the applicant is able to safely and efficiently conform to arrival and departure procedures and established traffic patterns at controlled and noncontrolled airports during day and night VFR operations.

A. Radio Communication and ATC Light Signals.

1. Description. The applicant may be asked to demonstrate

the use of designated frequencies and recommended voice procedures to report position and state intentions regarding the flight,
and to obtain pertinent information and clearances. Where applicable, the applicant is expected to use Airport Terminal Information
Service, Airport Advisory Service, Control Tower, Approach and
Departure Control, UNICOM, and ATC light signals.

2. Acceptable Performance Guidelines. The applicant shall
determine the type of communication facilities available, select
correct frequencies, and use appropriate communications procedures to obtain and acknowledge necessary information. Failing to
comply with airport traffic procedures or instructions without permission to do so shall be disqualifying.

B. Airport and Runway Markings and Lighting.

1. Description. Where available, the applicant may be asked
to demonstrate the proper use of wind and traffic direction indicators, and markings indicating closed runways, displaced thresh-
olds, taxiways, holding lines, and basic runways, and is also expected to be familiar with taxiway and runway lighting, rotating
beacons, obstruction lights, and VASI (Visual Approach Slope Indicator).

2. Acceptable Performance Guidelines. The applicant shall
know the meaning of standard wind and traffic indicators, markings
and lighting, and how they relate to airplane operation. Failure to
properly use these aids, creating an unsafe situation, shall be disqualifying.

C. Operations on the Surface.

1. Description. The applicant may be asked to demonstrate
safe operating practices while in close proximity to other aircraft,
persons, or obstructions. Emphasis should be placed on use of
brakes and power to control taxi speeds, proper positioning of flight
controls for existing wind conditions, awareness of possible ground
hazards, and compliance with taxi procedures and instructions. The
applicant is expected to take extra precautions when taxiing behind
large aircraft.

2. Acceptable Performance Guidelines. The applicant shall
maneuver the airplane on the surface without endangering persons
or property or conflicting with a smooth and orderly flow of traffic.

D. Traffic Patterns.

1. Description. The applicant may be asked to demonstrate
prescribed arrival and departure procedures and is expected to
maintain appropriate altitudes, airspeeds, and ground track consistent with instructions received or the established traffic pattern.

2. *Acceptable Performance Guidelines.* The applicant's performance shall be evaluated on the ability to maneuver the airplane relative to the runway in use. Consideration shall be given to application of wind drift corrections, adequate spacing in relation to other aircraft, and maintaining and controlling altitude and airspeed. Deviation of ±100 ft. from prescribed traffic pattern altitudes or ±10 knots from recommended airspeeds shall be considered disqualifying unless corrected promptly.

E. Collision Avoidance Precautions.

1. *Description.* The applicant is expected to exercise conscientious and continuous surveillance of the airspace in which the airplane is being operated to guard against potential mid-air collisions. Adequate clearing procedures should precede the execution of maneuvers involving rapid altitude and heading changes. The applicant shall perform whatever clearing is deemed necessary to ascertain that the area is clear before performing maneuvers such as stalls or flight at critically slow airspeeds, etc. There should be no delay in entering a maneuver upon completion of the clearing turn(s). This can be accomplished by performing the necessary conditions of flight (reducing airspeed, adding carburetor heat, etc.) while in the clearing turn(s). In addition to "see and avoid" practices, the applicant is expected to use VFR Advisory Service at nonradar facilities, Airport Advisory Service at nontower airports or FSS Locations, and Radar Traffic Information Service, where available.

2. *Acceptable Performance Guidelines.* The applicant shall maintain continuous vigilance for other aircraft and take immediate actions necessary to avoid any situation which could result in a mid-air collision. Extra precautions shall be taken, particularly in areas of congested traffic, to ensure that the view of other aircraft is not obstructed by the airplane"s structure. When traffic advisory service is used, the applicant shall understand terminology used by the radar controller in reporting positions of other aircraft. Failure to maintain proper surveillance shall be disqualifying.

F. Wake Turbulence Avoidance.

1. *Description.* The applicant may be asked to explain how, where, and when wingtip vortices are generated and their characteristics and associated hazards, and he should follow recommended courses of action to remain clear of these hazards.

2. *Acceptable Performance Guidelines.* The applicant shall identify the conditions and locations in which wingtip vortices may be encountered and adjust the flightpath so as to avoid these areas.

Failure to follow recommended procedures for avoiding the hazards of flying into wingtip vortices shall be disqualifying.

III. Flight Maneuvering by Reference to Ground Objects

Objective: To determine that the applicant can maneuver the airplane at approximately traffic pattern altitude over a predetermined groundpath while dividing attention inside and outside the airplane.

A. S-Turns Across a Road.

1. Description. The applicant may be asked to demonstrate a series of S-turns across a straight ground reference line approximately perpendicular to the wind. Bank variations should be planned to compensate for wind so that each half circle is equal on opposite sides of the line. At each reversal of direction, he should cross the selected line at a 90° angle with the wings level. A constant altitude should be maintained throughout the maneuver.

2. Acceptable Performance Guidelines. The applicant shall readily select ground references and maneuver the airplane in relation to these references. Properly coordinated turns, smooth control usage, and division of attention shall be required. Deviation of ±100 ft. from the selected altitude shall be considered disqualifying unless corrected promptly. Also, excessively steep banks, flight below minimum safe altitude prescribed by Regulations, or inadequate clearance of other aircraft shall be disqualifying.

B. Eights along a Road or Eights across a Road.

1. Description. The applicant may be asked to maneuver along a ground track starting above and parallel to a road, then perform 360° turns left and right. He is expected to vary the bank to correct for wind to arrive back over the road at the starting point upon completion of each 360° turn. The ground track should be in the form of a figure "8".

The applicant may be asked to perform a similar ground track maneuver starting over the intersection of two roads or some point on a road. The turns should be made so the intersection or point, which forms the center of the eight, is crossed in straight-and-level flight. A constant altitude should be maintained throughout the maneuver.

2. Acceptable Performance Guidelines. The applicant shall maneuver the airplane so the loops of the "eight" are symmetrical. Performance shall be evaluated on proper wind drift correction, airspeed control, coordination, altitude control, and vigilance for other aircraft. Deviation of ±100 ft. from the selected altitude shall be considered disqualifying unless corrected promptly. Also, excessively steep banks, flight below minimum safe altitude prescribed by Regulations, or inadequate clearance of other aircraft shall be disqualifying.

C. Rectangular Course.

1. Description. The aplicant may be asked to follow a rectangular or square course around and outside of a selected area. He is expected to correct for wind drift so the ground track is parallel to the sides of the selected area and equidistant from each side. A constant altitude should be maintained throughout the maneuver. This pattern should be performed both to the right and to the left.

2. Acceptable Performance Guidelines. The applicant shall readily select the ground reference and maintain the desired track in relation to that reference. Properly coordinated turns, smooth control usage, and division of attention shall be required. Deviation of ±100 ft. from the selected altitude shall be considered disqualifying unless corrected promptly. Also, excessive maneuvering to correct for wind drift, flight below minimum safe altitude prescribed by Regulations, or inadequate clearance from other aircraft shall be disqualifying.

D. Turns about a Point.

1. Description. The applicant may be asked to perform a ground track maneuver in which a constant radius of turn is maintained by varying the bank to compensate for wind drift, so as to circle and maintain a uniform distance from a prominent reference point on the ground. A constant altitude should be maintained throughout the maneuver. This maneuver should be performed both to the right and to the left.

2. Acceptable Performance Guidelines. The applicant shall maneuver the airplane so that the ground track is a constant distance from the reference point. Performance shall be evaluated on proper wind drift correction, airspeed control, coordination, altitude control, and vigilance for other aircraft. Deviation of more than ±100 ft. from the selected altitude shall be considered disqualifying unless corrected promptly. Also, excessively steep banks, flight below minimum safe altitude prescribed by Regulations, or inadequate clearance from other aircraft shall be disqualifying.

E. Eights around Pylons.

1. Description. The applicant may be requested to perform right and left turns around two ground reference points or pylons. A turn should be made in each direction, varying bank to correct for wind drift, resulting in a constant distance from each point. The ground track should be in the form of a figure eight.

2. Acceptable Performance Guidelines. The applicant shall maneuver the airplane so that both loops of the eight are of equal size. Performance shall be evaluated on proper wind drift correction, airspeed control, coordination, altitude control, and vigilance for other aircraft. Deviation of ±100 ft. from the selected altitude shall be considered disqualifying unless corrected promptly. Also, excessively steep banks, flight below minimum safe altitude prescribed by Regulations or inadequate clearance from other aircraft shall be disqualifying.

IV. Flight at Critically Slow Airspeeds

Objective: To determine that the applicant understands the reason for and can recognize changes in the airplane flight characteristics at critically slow airspeeds in various attitudes and configurations. To determine that the applicant can recognize imminent and full stalls and can accomplish prompt, positive, and effective recoveries in all normally anticipated situations.

A. Maneuvering at Minimum Controllable Airspeed.

1. Description. The applicant may be asked to maneuver in various configurations and at such airspeeds that controllability is minimized to the point that if the angle of attack is further increased by an increase in load factor or a decrease in airspeed, an immediate stall would result. The maneuver should be accomplished in straight flight, turns, climbs, and descents, using various flap settings (if applicable).

2. Acceptable Performance Guidelines. The applicant shall be evaluated on the ability to establish the minimum controllable airspeed, to positively control the airplane, to use proper torque corrections, and to recognize incipient stalls. Primary emphasis shall be placed on airspeed control. During straight-and-level flight at this speed, the applicant shall maintain altitude within ±100 ft. and heading within ±10° of that assigned by the examiner. Inadequate surveillance of the area prior to and during the maneuver or an applicant-induced unintentional stall shall be disqualifying.

B. Imminent Stalls.

1. Description. The applicant may be asked to demonstrate recoveries from imminent stalls entered from straight flight and from turning flight with power-on or power-off. He is expected to place the airplane in the attitude and configuration appropriate for flight situations such as takeoffs, departures, landing approaches, and accelerated maneuvers, as directed by the examiner. The applicant should apply control pressures which result in an increase in angle of attack until the first buffeting or decay of control effectiveness is noted. The recovery should be accomplished immediately by reducing the angle of attack with coordinated use of flight and power controls.

2. Acceptable Performance Guidelines. The applicant shall recognize the indications of an imminent stall and take prompt, positive control action to prevent a full stall. The applicant shall be disqualified if a full stall occurs or if it becomes necessary for the examiner to take control of the airplane to avoid excessive airspeed, excessive loss of altitude, or a spin.

C. Full Stalls.

1. Description. The applicant may be asked to demonstrate recoveries from full stalls entered from straight flight and from turning flight with power-on or power-off. He is expected to establish the attitude and configuration for flight situations such as takeoffs and departures, landing approaches, and accelerated maneuvers. Then, increase the angle of attack smoothly until a stall occurs, as indicated by a sudden loss of control effectiveness or uncontrollable pitching. Recovery should be accomplished by reducing the angle of attack immediately, and positively regaining normal flight attitude with coordinated use of flight and power controls. The applicant is expected to be aware of the loss of altitude necessary to recover from a stabilized high rate of descent with the elevator control fully back, if this condition is encountered before a stall develops.

2. Acceptable Performance Guidelines. The applicant shall recognize when the stall has occurred and take prompt action to prevent a prolonged stalled condition. The applicant shall be disqualified if a secondary stall occurs or if it becomes necessary for the examiner to take control of the airplane to avoid excessive airspeed, excessive loss of altitude, or a spin.

V. Takeoffs and Landings

Objective: To determine that the applicant can accomplish safe takeoffs and landings under all normally anticipated conditions in a landplane.

A. Normal and Crosswind Takeoffs.

1. Description. The applicant may be asked to demonstrate

normal and crosswind takeoffs by aligning the airplane with the runway or takeoff surface and applying takeoff power smoothly and positively while maintaining directional control. A pitch attitude should be established to permit normal acceleration to the manufacturer's recommended lift-off speed, and then smoothly increased to establish a straight climb at the desired climb speed.

The applicant may be asked to make at least one crosswind takeoff with sufficient crosswind to require the use of crosswind techniques, but not in excess of the crosswind limitations of the airplane used. In the crosswind takeoff, he is expected to hold aileron into the wind and maintain a straight path by use of rudder. These crosswind corrections should be maintained until after lift-off and the airplane then crabbed into the wind to prevent drift.

2. Acceptable Performance Guidelines. The applicant's performance of normal and crosswind takeoffs shall be evaluated on application of power and flight controls, directional control, coordination, and smoothness in establishing lift-off and climb. The applicant shall maintain a track aligned with the runway and a climb speed within ±5 knots of the desired climb speed after lift-off.

B. Normal and Crosswind Landings.

1. Description. The applicant may be asked to demonstrate normal and crosswind landings. Normal landings should be made using a final approach speed equal to 1.3 times the stalling speed in landing configuration (1.3 V_{so}), or the final approach speed prescribed by the manufacturer. Action should be taken to progressively reduce power so that the throttle is closed when the desired touchdown point is assured, or while rounding-out for touchdown. If the airplane is equipped with flaps, landings may be made with full flaps, partial flaps, or no flaps. Forward slips and a slip-to-a-landing may be performed with or without flaps, unless prohibited by the airplane's operating limitations.

In a tailwheel type airplane, the main wheels and tailwheel should touch the runway simultaneously at a near power-off stalling speed. In a nosewheel type airplane, the touchdown should be on the main wheels with little or no weight on the nosewheel. In strong, gusty surface winds, in a tailwheel type airplane, the round-out should be made to an attitude which permits touchdown on the main wheels only. Adequate corrections and positive directional control should be maintained during the after-landing roll.

The applicant may be asked to make at least one crosswind landing with sufficient crosswind to require the use of crosswind techniques, but not to exceed the crosswind limitations of the airplane. In crosswind conditions, wind drift corrections should be made throughout the final approach and touchdown.

The applicant may be asked to discontinue a landing approach at any point and execute a go-around.

2. Acceptable Performance Guidelines. The applicant's performance of normal and crosswind landings shall be evaluated on the basis of his landing technique, judgment, wind drift correction, coordination, power technique, and smoothness. Proper final approach speed shall be maintained within ±5 knots and touchdown in the proper landing attitude within the portion of the runway or landing area specified by the examiner.

Touching down with an excessive side load on the landing gear or poor directional control shall be disqualifying.

On go-arounds the applicant shall maintain positive airplane control, appropriate airspeeds, and operate the flaps and gear (if applicable) in proper sequence.

VI. Maneuvering by Reference to Instruments

Objective: To determine that the applicant is able to control and maneuver an airplane solely by reference to flight instruments as might be experienced under emergency conditions, and to use the emergency assistance available through radio aids, radar and DF (direction finding) heading instructions.

A. Basic Maneuvers.

1. Description. The applicant may be asked to demonstrate ability to control and maneuver the airplane solely by reference to flight instruments while performing straight-and-level flights, turns, climbs and descents, and while recovering from critical flight attitudes.

2. Acceptable Performance Guidelines. The applicant's performance shall be evaluated on coordination, smoothness, and accuracy. Turns shall be performed at least 180° to within ±20° of a preselected heading, and climbs and descents to within ±100 ft. of a preselected altitude. If the examiner finds it necessary to take over to avoid a stall or to avoid exceeding the operating limitations of the airplane, the applicant shall be disqualified.

B. Use of Radio Aids.

1. Description. Under simulated instrument conditions the applicant may be asked to follow a VOR radial or "home" to a radio station using ADF (Automatic Direction Finder), as appropriate to the radio equipment in the airplane. No prescribed orientation procedure will be required.

2. Acceptable Performance Guidelines. The applicant shall follow a radial or "home" to a station while effectively controlling altitude, heading, and airspeed.

C. Use of Radar or DF Heading Instructions.

1. Description. The applicant may be asked to demonstrate the proper procedures for contacting Approach Control or Flight Service Stations to request emergency assistance. He should be able to follow radar or DF heading instructions while in simulated instrument conditions.

2. Acceptable Performance Guidelines. The applicant's performance shall be evaluated on the ability to obtain and follow radar or DF heading instructions and emergency approach assistance received by radio, while effectively controlling altitude, heading, and airspeed.

VII. Cross-Country Flying

Objective: To determine that the applicant can prepare for and conduct a safe, expeditious cross-country flight.

A. Flight Planning.

1. Description. The applicant may be asked to plan a cross-country flight to a point at least 2 hours away at the cruising speed of the airplane used. At least one intermediate stop should be included. Planning should include the obtaining of pertinent and available weather information; plotting the course on an aeronautical chart; selecting checkpoints; measuring distances; and computing flight time, headings, and fuel requirements. The Airman's Information Manual should be used as a reference for airport information, NOTAMS, and such other appropriate guidance as may be extracted from its contents.

2. Acceptable Performance Guidelines. All flight planning operations shall be meaningful, accurate, and applicable to the trip proposed. The applicant shall explain the plan for the flight, verify the calculations, and present sources of information and data.

B. Conduct of Planned Flight.

1. Description. The applicant may be asked to perform the planned flight using pilotage, dead reckoning, and VOR or ADF radio aids as appropriate to the equipment in the airplane. He should make good the desired track, determine position by reference to landmarks, and calculate estimated times of arrival over checkpoints. He may also be asked to intercept and follow a VOR radial or "home" to a radio station using ADF, recognize station passage, and determine position by means of cross bearings.

The applicant should set out on the cross-country flight which was planned before takeoff. The planned course should be followed at least until the applicant establishes the compass heading necessary to stay on course, and can give a reasonable estimate of groundspeed and time of arrival at the first point of intended landing.

2. Acceptable Performance Guidelines. The applicant shall: (1) establish and maintain headings required to stay on course; (2) correctly identify position; (3) provide reasonable estimates of times of arrival over checkpoints and destination with an apparent error of not more than 10 minutes; and (4) maintain altitude within ±200 ft. of the planned altitude.

C. Diversion to an Alternate.

1. Description. When requested by the examiner to divert to an alternate airport, as might be necessary to avoid adverse weather, the applicant is expected to turn to the new course promptly. This may be accomplished by means of pilotage, dead reckoning, or radio navigation aids.

2. Acceptable Performance Guidelines. The applicant shall establish the appropriate heading for the course to the alternate and within a reasonable time give an acceptable estimate of the flying time and required fuel.

VIII. Maximum Performance Takeoffs and Landings

Objective: To determine that the applicant can use techniques appropriate to takeoffs and landings on short fields and on soft/rough fields.

A. Short Field Takeoff and Maximum Climb.

1. Description. The applicant may be asked to demonstrate a takeoff from a short takeoff area or over simulated obstacles. Power should be applied promptly and smoothly, and rotated to lift-off just as the best angle-of-climb airspeed is attained. The applicant is expected to maintain that speed until the assumed obstacles have been cleared. The applicant is expected to know and understand the effectiveness of the best angle-of-climb and the best rate-of-climb airspeeds of the airplane to obtain maximum climb performance. The flap settings and airspeeds prescribed by the airplane manufacturer should be used.

2. Acceptable Performance Guidelines. The applicant's per-

formance shall be evaluated on the basis of planning, smoothne̖,
directional control, and accuracy. The liftoff and climb shall be
performed within ±5 knots of the best angle-of-climb speed and the
assumed obstacle cleared by a safe margin.

B. Short Field Landing.

1. Description. The applicant may be asked to demonstrate a
landing from over an assumed 50-ft. obstacle. The final approach
speed should result in little or no floating after the throttle is closed
during the flare for touchdown. The airplane should clear the obstacle by a safe margin and touch down within the area designated by
the examiner, at minimum controllable airspeed. Upon touchdown
in landplanes, the applicant is expected to properly apply brakes or
reverse thrust to minimize the after-landing roll. Power, flaps or
moderate slips may be used as necessary on the last segment of the
final approach.

2. Acceptable Performance Guidelines. The applicant's performance shall be evaluated on planning, coordination, smoothness,
and accuracy. He shall control the angle of descent and airspeed on
final approach so that the assumed obstacle is safely cleared and
floating is minimized during the flare. After touchdown, he shall
bring the airplane smoothly to a stop within the shortest possible
distance consistent with safety.

C. Soft Field Takeoff.

1. Description. The applicant may be asked to demonstrate a
takeoff from a simulated soft field. This should be accomplished
with the wing at a relatively high angle of attack so as to transfer the
weight from the wheels to the wing as soon as possible. As soon as
the elevators become effective a positive angle of attack should be
established to lighten the load on the nosewheel or the tail-wheel.
After becoming airborne, the pitch attitude should be adjusted with
the wheels just clear of the surface to allow the airplane to accelerate. Care should be exercised to prevent settling back to the ground.
As the airplane reaches best angle-of-climb or best rate-of-climb
speed, whichever is appropriate for the field conditions, adjust the
pitch attitude to maintain the desired climb speed. The flap setting
used should be in accordance with the manufacturer's recommendations.

2. Acceptable Performance Guidelines. The applicant's performance shall be evaluated on planning, directional control,
smoothness, and accuracy. The applicant shall lift off at a speed not
higher than the power off stalling speed and maintain the proper
climb speed within ±5 knots.

Field Landing.

Description. The applicant may be asked to demonstrate a soft field landing from a normal approach with touchdown west possible airspeed to permit the softest possible touchdown and a short landing roll. A nose-high attitude should be maintained during the after-landing roll and the flaps promptly retracted (if recommended by the manufacturer) to prevent damage from mud and slush thrown by the wheels.

2. Acceptable Performance Guidelines. The applicant's performance shall be evaluated on planning, smoothness, and accuracy. He shall maintain the final approach airspeed within ±5 knots of that prescribed. *During flap retraction the applicant shall exercise extreme caution and maintain positive control.*

IX. Night Flying—Night VFR Navigation*

Objective: To determine that the applicant can properly prepare for a night flight and is thoroughly familiar with all aspects of night takeoffs and landings and night VFR cross-country flights.

A. Preparation and Equipment.

1. Description. The applicant may be asked to demonstrate how to prepare for a local or cross-country night flight. This requires familiarity with: (1) airport lighting; (2) the airplane's lighting system and its operation; (3) the need for a personal flashlight; and (4) the weather conditions pertinent to night flight. Particular attention should be given to the temperature/dewpoint spread due to the possibility of ground fog forming during night flights.

2. Acceptable Performance Guidelines. The applicant shall explain the significance of the items peculiar to the preparation for night flights.

B. Takeoffs and Landings.

1. Description. An actual demonstration of takeoffs and landings at night may be required. If required, the applicant is expected to explain and/or demonstrate: (1) proper use of power during the approach and landing phase; (2) efficient use of landing lights; (3) safe climb and approach paths; (4) safe taxi speeds; (5) recognition of position relative to other aircraft by the location and color of their lights; and (6) the dangers of spatial disorientation. If an actual demonstration is not required, the foregoing may be satisfied by oral quizzing.

2. Acceptable Performance Guidelines. The applicant's per-

---

*This pilot operation is not required if the applicant does not meet the night flying requirements set forth in Section 61.109. The certificate will bear the limitation, "Night Flying Prohibited".

formance shall be evaluated on ability to explain or demonstrate, as required by the examiner, the various techniques and aspects of night takeoffs and landings. Understanding the importance of constant vigilance for other aircraft on the ground and in the air, and the precautions necessary to avoid wake turbulence and spatial disorientation should be demonstrated.

C. VFR Navigation.

1. Description. An actual demonstration of night navigation may be required. If required, the applicant is expected to follow procedures similar to those described in this guide under "Cross-country Flying." If an actual night demonstration is not required, the foregoing may be satisfied by a daylight demonstration or oral quizzing.

2. Acceptable Performance Guidelines. The applicant's performance shall be evaluated on the Acceptable Performance Guidelines under "Cross-country Flying" in this guide, with special emphasis on the peculiarities of night flying.

X. Emergency Operations

Objective: To determine that the applicant can react promptly and correctly to emergencies which may occur during flight.

A. Partial or Complete Power Malfunctions.

1. Description. The applicant may be asked to demonstrate a knowledge of corrective actions for: (1) partial loss of power; (2) complete power failure; (3) rough engine; (4) carburetor or induction system ice; (5) fuel starvation; and (6) fire in the engine compartment. The examiner may, with no advance warning, reduce power to simulate engine malfunction.

2. Acceptable Performance Guidelines. Performance shall be evaluated on the applicant's prompt analysis of the situation and the remedial course of action. The emergency procedures shall be performed in compliance with the manufacturer's published recommendations. Any action which creates unnecessary additional hazards shall be disqualifying.

B. Systems or Equipment Malfunctions.

1. Description. The applicant may be asked to demonstrate a knowledge of corrective actions for: (1) inoperative electrical system (generator, alternator, battery or circuit breaker); (2) electrical fire or smoke in cockpit; (3) gear or flap malfunctions; (4) door opening in flight; and (5) inoperative elevator trim tab. Where practicable, the examiner may, with no advance warning, simulate flap malfunctions, landing gear malfunctions, or an inoperative electrical system. The applicant is expected to actually perform the recommended emergency procedures for the simulated malfunction.

2. Acceptable Performance Guidelines. Performance shall be evaluated on the applicant's prompt analysis of the situation and his remedial course of action. The emergency procedure shall be in compliance with the manufacturer's published recommendations. Any action which creates unnecessary additional hazards shall be disqualifying.

C. Lost Procedures.

1. Description. The applicant may be asked to explain the proper courses of action to be taken if becoming lost, being trapped on top of an overcast, losing radio communications, or encountering unanticipated adverse weather.

2. Acceptable Performance Guidelines. Performance shall be evaluated on the applicant's ability to promptly and correctly analyze the situation and describe the appropriate remedial action.

XI. Applicant's Flight Test Checklist

In addition to a clean aircraft and any supporting material the flight instructor has suggested, the following is a typical checklist for your FAA flight test:

Acceptable Airplane with Dual Controls.

☐ View-Limiting Device (Hood).
☐ Aircraft Documents:
  Airworthiness Certificate.
  Registration Certificate.
  Operating Limitations.
☐ Aircraft Maintenance Records:
  Airworthiness Inspections.
☐ FCC Station License.
☐ Weight and Balance Data.

Personal Equipment.

☐ Current Aeronautical Charts.
☐ Computer and Plotter.
☐ Flight Plan Form.
☐ Flight Logs.

Personal Records.

☐ Pilot Certificate.
☐ Medical Certificate.

- ☐ Completed Application for an Airman Certificate And/Or Rating (FAA Form 8420-3).
- ☐ Airman Written Test-Report (AC Form 8060-37).
- ☐ Logbook with Instructor's Endorsement.
- ☐ Notice of Disapproval (if applicable).
- ☐ Approved School Graduation Certificate (if applicable).
- ☐ FCC Radiotelephone Operator Permit.
- ☐ Examiner's Fee (if applicable).

SUMMARY

In closing Part 3, simply remember that the flight examiner is just as anxious for you to pass the examination as you are. In all reality, you will find that not only are flight examiners very human (after all), but many provide instruction during the flight examination. One very important fact to remember is that your flight instructor is *also* being examined by way of your performance. A task the flight examiner must perform is to test you in all of the various maneuvers and flight operations to assure that your flight instructor has provided you a complete training program as well as your having mastered such.

Regarding the aircraft, be certain it is clean inside, with absolutely no loose items that may fly about! Have a weight and balance form made out in advance for the flight and offer it to the examiner before takeoff. Do not under any circumstances expect the examiner to fly in an aircraft that is out of balance, overweight, or lacking the proper records. Establish immediately that the test aircraft is fully compliant with the FARs and you will have already gotten to first base on your flight examination. A thorough preflight, radio test, and engine runup completes second base. Simply performing the maneuvers with "ordinary" skill is all that is needed to make home plate.

One very important point to remember is that you are pilot-in-command during the FAA flight test, *not the examiner!* The examiner is simply riding along to observe your performance. In the event the examiner should ask for a maneuver which is beyond the aircraft capabilities or in conflict with applicable FARs, it is *your* responsibility as pilot-in-command to so advise the examiner accordingly and *not* perform the maneuver. Can this happen? Of course—one cannot expect the examiner to remember all of the details and restrictions for all of the aircraft he encounters in his work. It is *your job* to know the restrictions of the test aircraft you have selected. Can the aircraft be slipped with full flaps? Is takeoff and climb-out to be made on one or both tanks? Is the aircraft certificated for spins? Help yourself (and the examiner if necessary) by thoroughly knowing your aircraft and its performance limits.

As well as testing your aeronautical knowledge, an examiner may quite logically examine your reaction to stress. This is accomplished by simply loading you up with tasks faster than you can perform them. In such an instance it is okay to tell the examiner to "stand by" while you are getting caught up, then advise him that you are ready for further work. He will wait. After all, he is just trying to determine your work limit and the manner in which you treat an overload. In short, it is not "I can't," it is "stand by."

In the event a maneuver goes poorly, simply request an opportunity to repeat the exercise. Don't make excuses, simply express the fact that you can do better and proceed to demonstrate so on a second attempt. After all, the FARs simply state that the candidate must show that he or she is master of the aircraft, with the successful outcome of a procedure or maneuver never seriously in doubt (FAR 61.43). If you have prepared well and have insisted on quality flight instruction, there will be no doubt. Success is the result of preparation and practice. It is not an accidental event.

Chapter 9
An Instrument/Commercial
Pilot Training Program

Just as in the case of the private pilot certificate, the instrument/ commercial candidate must know exactly where his training program is taking him. At today's rates, the cost of a commercial certificate can well exceed the price of a new car. With this thought in mind, the instrument/commercial student will get the most for his or her dollars by thoroughly understanding the course curriculum and objectives. Again, it is suggested that the student use the curriculum as a check list, noting those items that have been completed as the course progresses. By so doing, possible omissions of material become evident.

The order in which material is presented in a curriculum is not a primary consideration. Rather, it represents an idealized situation. One cannot instruct in crosswind landings on a calm day. Neither can ILS approaches be practiced when some other approach is in use, or approach control is too busy to accept practice approaches. A curriculum must be flexible and allow for the substitution of alternate activities when the occasion demands.

Surprisingly, the FAA does not publish an instrument/ commercial curriculum. Rather, there are simply curriculum examples as shown by FAA publications such as the *Instrument Flying Handbook* AC 61-27B. The curriculum that follows is one that I have used. It does embody those items noted in the *Instrument Flying Handbook* but is not to be construed as an FAA generated or approved curricula. It is intended to be used by a student or instructor to ensure that all FAA-required subjects have been treated in the training program. For simplicity, emphasis is placed principally on single engine landplane training. As was the case in Chapter 8, the subject matter is broken into parts as follows:

Part 1: A Ground School Curriculum.
Part 2: Flight Instruction Curriculum.
Part 3: The Instrument Flight Test.
Part 4: The Commercial Pilot Flight Test.

In order to fit either the instrument student or the commercial student, the material for such has been separated accordingly. In reality, the commercial rating is nothing more than an instrument rating plus required commercial flight maneuvers and a Class II physical. The commercial may thus be looked upon as simply an extension of the instrument rating. Since many students obtain their instrument rating first and follow with the commercial, the curriculum of this text has been arranged accordingly.

For those students who desire their commercial certificate but haven't their instrument rating, the following table illustrates an arrangement of flying lessons for blending both the instrument and commercial maneuvers into a single course. Lesson numbers refer to those listed in Part 2 of this chapter. As will be noted, a number of the commercial lessons provide preparation for their more difficult instrument counterparts.

A Typical Commercial Lesson Sequence

Instrument Lessons   *Commercial Lessons*

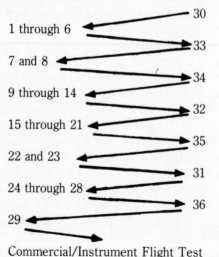

Commercial/Instrument Flight Test

In closing this introduction it has been assumed that the reader of this chapter is already a private pilot with some degree of experience (single or multi-engine) who is interested in obtaining an advanced rating. For this reason, certain parts of the lesson plans are somewhat less detailed than those contained in Chapter 8. Where relevant, sample approach plates, SIDs and STARs are included as supporting illustrations. For a complete explanation thereof, the student is referred to a text on instrument flying such as the FAA publication *Instrument Flying Handbook*, AC 61-27B.

As an aid to estimating costs, typical figures for instruction and solo hours have been noted for each of the blocks of learning. These figures are based upon a private pilot candidate with 150 to 200 hours of experience in a simple aircraft. Naturally, hour estimates will vary considerably, depending upon total experience as well as recency of experience. In addition, previous time in a complex aircraft and/or the availability of an IFR trainer can significantly affect the time it will require to complete an individual course of instruction. Minimize your time and costs by understanding course requirements as outlined in the curriculum.

PART I: A GROUND SCHOOL CURRICULUM

Refer to the Pilot/Controller Glossary and a Basic IFR Clearance Shorthand for a definition of terms used in this chapter.

Lesson No. 1—Federal Aviation Regulations

Objective. To develop the student's knowledge and understanding of Federal Aviation Regulations (FARs) and our system of aviation government.
Content.
1. Structure of our aviation government.
2. How FARs come into being.
3. The role of the National Transportation and Safety Board (NTSB).
4. Federal Communications Commission (FCC) requirements.
5. FAR Part 1, Definitions and Abbreviations.
6. FAR Part 61, Certification of Pilots and Flight Instructors.
7. FAR Part 71, Airspace Designations.
8. FAR Part 91, General Operating Flight Rules.
9. FAR Part 97, Standard Instrument Approach Procedures.
10. NTSB Part 830, Aircraft Accidents, Overdue Aircraft, and Safety Investigations.

11. Supporting material:
    a. Airman's Information Manual (AIM).
    b. Required aircraft documents.
    c. Advisory Circulars.
    d. Exam-O-Grams.
    e. NTSB safety recommendations.

Completion Standards. The lesson will be successfully completed when the student passes a written quiz on FAR Parts 1, 61, 91 and NTSB Part 830. Estimated classroom instruction time: 6 hours.

Lesson No. 2 — Aeromedical Considerations

Objective. To develop the student's knowledge of aeromedical facts, particularly as they relate to IFR flight.

Content.
1. The nature of vertigo.
2. Hypoxia, the use of oxygen.
3. Hyperventilation.
4. Bends, scuba diving precautions.
5. CO poisoning, recovery time.
6. Alcohol and drugs, synergistic reactions.
7. Ear-sinus blockage.
8. Noise and hearing damage; earplugs and headsets.
9. Fatigue.
10. Night flight:
    a. Vision and the use of oxygen.
    b. Mathematical reasoning and hypoxia.
    c. Colorblindness.
    d. Flicker vertigo.
    e. Optical illusions (flying on a star).
11. In-flight emergencies:
    a. Pilot or passenger becomes ill.
    b. Emergency authority of the pilot-in-command.
    c. "Life Guard" flights.
12. USAF high altitude "pressure tank" training. Discuss and/or recommend training for students who may be able to visit a USAF or Navy facility that offers such to civilians.

Completion Standards. The lesson will be successfully completed when the student passes appropriate oral quizzing on the subject of aviation medical considerations. Estimated classroom instruction time: 2 hours.

Lesson No. 3 — Aerodynamic Forces

Objective. To review and develop the student's knowledge of aerodynamic forces related to commercial VFR and IFR flight.

Content.

1. Axes of motion.
2. Four basic forces: ⎫ Brief review
   a. Lift vs. weight. ⎬
   b. Thrust vs. drag. ⎭
3. Lift "equation" parameters:
   a. Wing camber/flaps.
   b. Angle of attack:
      Coeffecient of lift.
      Coefficient of drag.
   c. Stalls and stall control:
      Airfoil shape/wing twist.
      Stall strips/slots/slats.
      Wing planform.
   d. Wing area/wing loading/load factor.
   e. Velocity squared considerations:
      Bernoulli theorem.
      Verturi.
   f. Air density/density altitude.
4. Adverse yaw.
5. Operation of the trim table
6. Stability in flight:
   a. Weight and balance.
   b. Dihedral.
   c. Pendulum effect.
   d. Wing sweepback.
   e. Dorsal fin (spin stability).
7. Propeller considerations:
   a. P-factor.
   b. Torque.
   c. Corkscrew slipstream.
   d. Gyroscopic precession.
8. Flight optimums, how derived:
   a. Best rate of climb.
   b. Best angle of climb.
   c. Best endurance (glide and cruise).
   d. Best range (glide and cruise).
9. Maneuver envelope. ⎫ Brief review
10. Gust envelope ⎬

Completion Standards. The lesson will be successfully completed when the student satisfactorily answers oral questions on the subject of aerodynamics. Estimated classroom instruction time: 3 hours.

Lesson No. 4—Aircraft Performance Limits

Objective. To further develop the student's understanding of aircraft performance limits and methods for calculating important parameters.

Content.
1. Air density and performance:
   a. Engine horsepower.
   b. Take off climb.
   c. Rate of climb.
   d. Stall speed.
   e. Ceiling.
   f. Calculation of density altitude.
   g. De Nault computer.
2. Cruise performance:
   a. Fixed-pitch propeller aircraft.
   b. Constant-speed propeller aircraft.
3. Weight and balance:
   a. Graph method of computation.
   b. Table method of computation.
4. Crosswind component chart.
5. Related subjects:
   a. Soft field takeoffs.
   b. Effect of humidity on horsepower.
6. Emergency situations:
   a. Stuck flaps.
   b. Power surges.
   c. Landing gear malfunctions.

Completion Standards. The lesson will be completed when the student successfully accomplishes an example cross-country flight planning problem involving lesson principles. Estimated classroom instruction time: 3 hours.

Lesson No. 5—Aircraft Engines

Objective. To refresh the student's understanding of aircraft engine performance limits and the operation of related engine instrumentation.

Content.
1. Brief review—aircraft engine operating principles and construction.
2. Propeller operation—review.
3. Basic engine operating limitations:
   a. Start and warm-up (cold weather preheat).
   b. Rpm, prohibited regions.
   c. Takeoff and cruise ratings (leaning).
   d. Cowl flap operation, engine temperature control.
   e. Overboost (turbo engines).
4. Aviation gasoline:
   a. Color code, octane ratings, detonation.
   b. Why auto gas should not be used.
5. Aviation oils:
   a. Types
   b. Why additives should not be used.
6. Engine subsystems:
   a. Carburetor operation, icing.
   b. Fuel injection systems.
   c. Magneto and spark plugs.
   d. Alternator/generator and battery system.
   e. Oil flow and cooling.
7. Engine instruments:
   a. Manifold pressure and rpm, relation to horsepower.
   b. Suction gauge and pump.
   c. Oil temperature and pressure.
   d. Fuel pressure, flow, and quantity.
   e. Ammeter and voltmeter.
   f. Cylinder head temperature.
   g. Carburetor throat temperature.
   h. Exhaust gas temperature.
8. Engine maintenance:
   a. Spark plug fouling.
   b. Time between overhauls (TBO).
   c. Overhaul vs. remanufacture.
   d. Engine storage (short term/long term).

Completion Standards. This lesson will be completed when the student satisfactorily answers oral questions on the elements of the lesson. Estimated classroom instruction time: 3 hours.

Lesson No. 6—Navigation Instruments

Objective. To refresh and reinforce the student's understanding of the basic navigation and aircraft attitude control instruments, with emphasis on IFR flight.

Content.

1. Pitot-static system:
   a. Altimeter.
   b. Vertical speed indicator.
   c. Airspeed indicator.
2. Gyro system:
   a. Altitude indicator.
   b. Directional gyro.
   c. Turn coordinator.
   d. Needle and ball.
3. Magnetic Compass.

} Operating principles and instrument errors.

4. Complex aircraft instrumentation:
   a. Horizontal situation indicator (HSI).
   b. Radio magnetic indicator (RMI).
   c. Flight director.
5. IFR Instrument relationships:
   a. Pitch indicators.
   b. Turn indicators.
   c. Primary-supporting instruments.
   d. Flying on primary panel.
6. Instrument maintenance and test:
   a. Effect of smoking on gyro life.
   b. Test of magnetic compass by reference to the directional gyro.
   c. Test of turn coordinator by timed turns.

Completion Standards. This lesson will be successfully completed when, by oral test, the student displays a working knowledge of instrument operating principles and instrument errors. Estimated classroom instruction time: 2 hours.

Lesson No. 7 — Radio/Radar Navigation

Objective. To reinforce the student's knowledge of Radio and Radar Navigation with emphasis on IFR flight.

Content.

1. Radio propagation characteristics:
   a. Low frequencies-storms, night effect, fading.
   b. High frequencies-line of sight restriction.
2. Communications facilities:
   a. Flight Service Station (FSS).

    b. Automatic Terminal Information Service (ATIS).

    c. Transcribed Weather Broadcast (TWEB).

    d. Clearance Delivery. ⎫

    e. Ground Control. ⎪

    f. Tower. ⎬ _____ Air Traffic Control

    g. Departure Control. ⎪ (ATC).

    h. Center. ⎪

    i. Approach Control. ⎭

    j. Unicom and air-to-air communications.

    k. Emergency frequency.

3. Radio navigation facilities and operation:

    a. Omni group:

            VOR, Very hi frequency Omni Range.

            DME, Distance Measuring Equipment.

            RNAV, Area Navigation.

            VOT, Very hi frequency Omni Test.

    b. Instrument Landing System group:

            Localizer

            GS, Glide Slope.

            OM, Outer Marker.

            MM, Middle Marker.

            IM, Inner Marker.

            FM, Fan Marker.

    c. Automatic Direction Finder group:

            NDB, Non Directional Beacon.

            LOM, Locator Outer Marker.

            Broadcast station.

4. Radar navigation facilities and operation:

    a. Radar principles:

            Primary radar-compare accuracy to VOR/DME.

            Secondary radar-transponder, operating modes.

    b. Radar types:

            ASR, Airport Surveillance Radar.

            PAR, Precision Approach Radar.

            ARSR, Air Route Surveillance Radar.

    c. Radar services:

            VFR, Stages I, II, and III.

            IFR, Radar vectoring.

            MSAW, Minimum Safe Altitude Warning.

            CA, Collision Avoidance.

            ARTS III, Automated Radar Terminal System.

Limited ATC storm warning capability.

Emergency services.

5. Related Facilities:

   a. Weather bureau storm warning radar.

   b. Aircraft weather radar.

   c. Stormscope.

   d. Microwave ILS (Air Carriers).

Completion Standards. This lesson will be successfully completed when, by oral test, the student displays a working knowledge of radio/radar navigation as applied to an IFR cross-country class discussion problem. Estimated classroom instruction time: 3 hours.

Lesson No. 8 — Weather?

Objective. To reinforce the student's general understanding of weather and develop the student's ability to assess weather conditions pertaining to IFR flight.

Content—Part A.

1. Reason for study of weather systems—NTSB weather related accident statistics.

2. Global weather forces:

   a. Non uniform heating of the atmosphere, global flow patterns, standing hi and lo pressure regions.

   b. Cyclonic flow.

   c. Coriolis effect-friction layer.

   d. Jet stream-clear air turbulence (CAT).

   e. Pressure gradients.

   f. Thermals—vertical air flow.

3. Air masses and fronts—types and characteristics, inversions.

4. Clouds-their formation:

   a. Definition of standard atmosphere.

   b. Definition of atmospheric stability.

   c. Definition of supercooling.

   d. Cloud names and meanings.

   e. Flight characteristics—icing, turbulence.

5. Severe weather:

   a. Thunderstorms and squall lines.

   b. Tornadoes and hurricanes.

   c. Lightning, its effect on aircraft.

Content—Part B.

6. How weather is reported and forecast; the pilot's role as an observer:

   a. Weather bureau, weather radar.

b. Flight Service Station, EFAS.

c. TWEB, Transcribed Weather Broadcast.

d. PATWAS, Transcribed Telephone Service.

e. ATC/ARTCC, en route thunderstorm reports.

7. TWX Reports:

a. SA, Surface Aviation

b. FA, Forecast Area.

c. FT. Forecast Terminal.

d. FD, Forecast Winds Aloft. } Discuss contents and meaning

e. VA, Pilot Report (PIREP). } of reports

f. Airmet.

j. Sigmet.

8. Weather Charts:

a. SA, Surface Area.

b. Weather depiction.

c. Prognostic. } Discuss contents and meaning

d. Radar. } of chart features.

e. Satellites.

f. Hi altitude.

9. Weather forecasting accuracy:

a. Practical lightplane IFR weather limits.

b. Planning a way out.

c. ATC weather avoidance assistance.

d. Emergency DF approach.

10. Future computer-aided self briefing weather terminals.

Completion Standards. This very important lesson will be successfully completed when the student passes a comprehensive quiz on the reading and interpretation of TWX reports and weather charts. A visit to FSS is highly recommended. Estimated classroom instruction time: 6 hours.

Lesson No. 9— VFR Flight Planning

Objective. The objective of this lesson is to illustrate to the student the manner in which VFR flying practices blend with IFR flight.

Content.

1. VFR chart review:

a. Planning chart.

b. World Aeronautical Chart (WAC). } Pilotage and dead

c. Local chart. } reckoning review.

2. Sectional charts and IFR flight:
   a. Obstacles on the takeoff route: tower or departure control "forgets" or is unable to communicate with your.
   b. Obstacles off airways: while being radar vectored ATC "forgets" or is unable to communicate with you. You are descended to an MVA; as pilot-in-command is the assigned altitude acceptable?
3. Flight computer review:
   a. Time-speed-distance problems.
   b. Fuel consumption problems.
   c. Nautical/statute mile conversions.
   d. True airspeed/density altitude problems.
   e. True altitude problems.
   f. Drift correction triangles.
   g. Wind/ground speed problems.
   h. Finding the wind in flight.
   i. Mach number.
4. Airport lighting at night:
   a. IFR approach light systems.
   b. Runway and runway-end identifier lights (REIL).
   c. Taxiway and taxiway centerline lighting.
   d. Obstacle lighting.
   e. Rotating beacon.
   f. Tower control of lighting.
   g. Unattended airports, radio control of lighting.

Completion Standards. This lesson is successfully completed when the student demonstrates the ability to solve classroom assigned flight computer problems. Estimated classroom instruction time: 2 hours.

Lesson No. 10 — IFR Flight Planning

Objective: To acquaint the student with the various charts used in IFR flight and their features.

Content.
1. IFR Charts—NOAA and Jeppesen:
   a. Planning chart.
   b. En route—low-altitude.
   c. En route—/high altitude. } Discuss symbols and features.
   d. Local Area.
2. Approach/Departure Charts-NOAA and Jeppesen:
   a. STAR, Standard Terminal Arrival Route.

b. SID, Standard Instrument Departure.

c. Approach Plates, types and features.

d. Taxi Charts.

3. Compare NOAA and Jeppesen chart features.

4. Relate aircraft performance to chart climb/descent constraints.

Completion Standards. This lesson is complete when, by oral quizzing, the student demonstrates a thorough knowledge of en route charts, SIDs, STARs, and approach plates. Estimated classroom instruction time: 4 hours.

Lesson No. 11—Flying IFR

Objective. The objective of this lesson is to employ the knowledge gained in lessons 7, 9, and 10 in the conduct of example IFR procedures and cross-country flights. This lesson is intended to provide a *thorough* understanding of IFR navigation and procedure technique including related communications with ATC.

Content—Part A.

1. Preparation for IFR flight:

    a. IFR flight plan, fuel requirements, alternate airport.

    b. IFR checklist for aircraft.

    c. IFR charts, sectional charts.

    d. Night flight equipment.

2. Clearances:

    a. General clearance message construction.

    b. Use of clearance shorthand (see Appendix B).

    c. Pre-taxi clearance/readback.

    d. En route clearances/readback.

    e. Amending a clearance.

    f. Cleared for the approach, discuss meaning and minimum altitude restrictions.

    g. Refusing a clearance—request for alternate.

        Possible wake turbulence. ⎫

        Over ocean flight. ⎬ Example reasons.

        Thunderstorm avoidance. ⎭

    h. Clearance void time-departing a non-IFR airport.

    i. Cruise clearance.

    j. Tower-to-tower clearances.

    k. Cancelling an IFR flight plan.

3. Elements of en route IFR flight:

    a. MEA/MOCA definition and flight procedures:

        MRA, Minimum reception altitude.

MCA, Minimum crossing altitude.
Altitude change points.
b. Reporting points and procedures:
Mandatory and non-mandatory reports.
When in radar contact.
c. IFR on top.
d. Feeder routes.
e. SID/STAR routes, climb requirements.
f. RNAV routes.
g. Holding patterns, standard and non-standard:
At a VOR.
At an airway intersection. } Entry method
At a DME fix point. } Correction for wind.

Content—Part B.
1. Elements of IFR approaches.
   a. Meaning of "cleared for the approach".
   b. Segments of the approach.
   c. Procedure turn types.
   d. Delaying radar vectors:
      Expect approach time.
      Route of delaying path.
   e. Exact meanings and flight procedures relating to:
      MVA, Minimum Vectoring Altitude.
      MSA, Minimum Safe Altitude.
      FAF, Final Approach Fix.
      MAP, Missed Approach Point.
      DH, Decision Height.
      Missed Approach.
   f. Landing:
      Straight-in approach.
      Circling approach.
      Instrument runway markings.
      CAT II approach-brief review of requirements.
2. Flying the approach:
   a. VOR at destination airport.
   b. VOR not at destination airport.
   c. ILS approach, false slopes.
   d. Back course localizer approach.
   e. NDB at destination airport.
   f. NDB not at destination airport.
   g. DME arc/VOR or localizer.

h. RNAV approach.
i. ASR, Airport Surveillance Radar approach.
j. PAR, Precision Approach Radar approach.
k. Emergency DF approach.
l. Visual/Contact approach.
m. In discussing the subject of approaches, provisions for including the many types of approach features must be included. Examples are as follows:
    Combination VOR-ADF fix points.
    Step-down approach.
    Side-step maneuver.
    Approaches to non-IFR airports.
3. Emergency considerations:
    a. Loss of radio communications.
    b. Loss of aircraft suction system.
    c. Loss of aircraft electrical system.
    d. Loss of altitude gyro. ⎫
    e. Loss of directional gyro. ⎭ Partial panel operation.
    f. Partial loss of airport navaids, substitution of alternate navaids, increase in landing minimum limits.
    g. Weather below minimums at MAP or DH.
    h. Icing or CAT during the approach.

Completion Standards. This lesson is a key lesson in the process of learning IFR flight. To be successfully completed, the student must demonstrate knowledge of the subject by passing a comprehensive written quiz on the lesson material. A visit to a local ATC radar facility is highly recommended to reinforce this and related lesson material pertaining to IFR procedures. Estimated classroom instruction time: 6 hours.

Lesson No. 12—Commercial Flight Maneuvers (Commercial Students)

Objective. To develop the student's understanding of flight maneuvers required for the commercial certificate.
Content.
1. Review of common maneuvers:
    a. Flight at minimum controllable airspeeds.
    b. Soft field takeoffs and landings.
    c. Short field takeoffs and landings.
    d. Cross wind takeoffs and landings.
    e. Emergency landings.

2. Advanced maneuvers:
   a. Steep turns (60° bank angles).
   b. Chandells.
   c. Lazy Eights.
   d. Spiral descents.
   e. Accuracy landings.
3. Optional maneuvers (not required by FAR):
   a. Pylon turns.
   b. Pylon Eights. } Pivotal altitude.
   c. Spins/spin recovery.

Completion Standards. This lesson is complete when, by oral quizzing, the student demonstrates a knowledge of the principles involved in executing commercial maneuvers. Estimated classroom instruction time: 2 hours.

Lesson No. 13 — Complex Aircraft Systems (Commercial Students)

Objective. To develop the student's knowledge of the more complex equipment as typically found on high-performance single or twin-engine aircraft.

Content.
1. Turbocharging, general theory:
   a. Manual.
   b. Automatic. } Wastegate control systems.
   c. Fixed.
   d. Oil requirements, failure modes.
   e. Spool up/down times.
   f. Altitude limitations, cylinder head temps.
   g. Overboost limits and precautions.
   h. Turbo/Non-turbo engine differences.
2. Oxygen systems, permanent and portable:
   a. Passenger capacity vs. operating time.
   b. Mask types and oxygen flow systems.
3. Autopilots:
   a. Single, dual, and three-axis types.
   b. Navigation couplers.
   c. Altitude coupler.
4. Constant speed propellers:
   a. Twin engine rpm synchronizaton.
   b. Fuel feathering operation.
   c. Runaway propeller governor.
5. Retractable landing gear:
   a. Electromechanical type.

b. Hydraulic type.

c. Emergency operation.

6. Anti-icing/deicing systems.

7. Twin engine aircraft:

a. Critical engine.

b. Single engine operation, V_{MC} and V_{SSE}.

8. FAR Part 135, Air Taxi Operators and Commercial Operators of Small Aircraft; review of pilot requirements.

Completion Standards. This lesson is complete when, by oral quizzing, the student demonstrates a knowledge of complex aircraft systems and their operation. Estimated classroom instruction time: 3 hours.

Lesson No. 14 — Emergency Procedures

Objective. To review and refresh the student's knowledge of general aircraft emergency procedures.

Content.

1. Fire:

a. During engine start (overpriming).

b. In flight-gasoline/oil.

c. In flight—electrical.

2. Loss of oil; corrective actions:

a. Low oil pressure.

b. High oil temperature.

3. Contaminated fuel, corrective actions:

a. Water contamination.

b. Jet fuel contamination.

c. Improper octane contamination.

4. Low on fuel:

a. Best range power settings and relationship to wind.

b. Best endurance power settings.

c. Use of pilot-in-command emergency authority to obtain a landing priority.

d. Optimum method for running tanks dry.

5. Loss of alternator or generator:

a. Ground operations—auxiliary power unit engine start.

b. In flight—day.

c. In flight—night.

d. In flight—IFR conditions.

6. Alternator/Generator regulator failure:

a. Over voltage—possible battery damage.

b. Under voltage—improper battery charge.

7. Fuses and circuit breakers—age fatigue, improper operation.
8. Radio failure, use of squawk code 7600.
9. Severe weather conditions:
   a. Heavy icing, hail, snow.
   b. Frozen controls due to ice.
   c. Severe turbulence.
   d. Heavy rain.
   e. Lightning.
10. Loss of power:
    a. Cockpit check, engine restart attempt.
    b. Position report on 121.5 mhz, transponder set to 7700.
    c. Landing site selection.
    d. Emergency locator transmitter-armed or on.
    e. Best landing gear position (retractables).
    f. Orderly shutdown of gasoline and electrical equipment before touchdown.
    g. Ditching procedures-overwater flight.
11. Emergency supplies:
    a. Cold weather—snow.
    b. Hot weather—desert. } Discussion of emergency equipment.
    c. Over water.

Completion Standards. This lesson is complete when the student demonstrates a general knowledge of the subject in response to classroom discussions related to emergency procedures. Estimated classroom instruction time: 2 hours.

Lesson No. 15 — Course Review

Objective. To refresh and reinforce the student's knowledge of IFR flight through review and practical examples thereof.
Content.
1. General review. The instructor shall conduct a general course review with emphasis on those subjects which have been the more difficult as judged from student written and oral quizzes. The review shall include basic FAR's as related to IFR flight.
2. Cross country flight-class participation problem. By using an extended IFR cross-country problem as a review theme, the instructor shall conduct a guided discussion as the class "flies" the route. As a minimum, the cross-country problem shall provide for the following:
   a. Takeoffs and landings at 3 major airports which are at least 200 NM apart.

b. Time and fuel estimates.

c. Preparation of flight plans.

d. ATC clearances (the instructor).

e. An encounter with unforecast weather.

f. Arrival at a destination which is below minimums.

g. A microphone relay failure (can receive but not transmit).

h. En route clearance changes in course by ATC (the instructor).

i. Failure of the aircraft suction pump.

The problem shall be supported by applicable charts, SIDs, STARs, and approach plates as needed to "fly" the assigned route.

Completion Standards. The review lesson will be satisfactorily completed when the student(s) demonstrate a general knowledge of the subject by means of classroom discussion and quizzing. Estimated classroom instruction time: 3 hours.

Lesson No. 16—Example Final Examination

Objective. To evaluate the student' general knowledge and understanding of the course material.

Content. The example final examination shall consist of at least 100 questions which cover the content of the course. If the instructor so chooses, test questions may be selected from those as published by the government relative to the instrument or commercial written examination, whichever is appropriate to the class.

Completion Standards. The student shall demonstrate satisfactory knowledge of the course by a passing grade of 80% or better. Estimated classroom test and critique time: 4 hours.

Concluding Note

The preceding 16 lessons (blocks of learning) are estimated as requiring a minimum of 54 classroom hours. Depending upon student backgrounds, classroom facilities, and the availability of supporting visual aids, a classroom instruction time of up to 60 hours may be needed to present the material suggested. The instrument/ commercial course is one of the most complex courses to be mastered by a pilot. It should be taught by a seasoned instructor and accepted by the student as a serious learning endeavor. IFR flight is simply not forgiving of carelessness or neglect. Commercial flight carries with it the obligation of protecting the safety of the passenger. Stated rather directly, maturity and the motivation to conquer the difficult are required of the student who undertakes the Instrument/Commercial certificate.

PART 2: A FLIGHT INSTRUCTION CURRICULUM

Lesson No. 1 — Introduction to IFR Instrumentation

Objective. To promote familiarity with IFR related cockpit instruments and radio equipment.

Content—Part A.

During Part A of this lesson the student shall be introduced to preflight procedures which shall apply to *all* subsequent IFR lessons. The list that follows names general IFR checklist items. To this list, of course, must be added the standard list of cockpit and preflight walk-around items that apply to the aircraft being used.

1. IFR Preflight walk-around check items:
   a. Static port—clear.
   b. Pitot port—clear.
   c. Pitot heater—check for operation.
   d. Stall warning—check for operation.
   e. Stall warning heater—check for operation if applicable.
   f. Antennas—check.
   g. Vertical speed indicator—check/set zero position.
2. Cockpit check prior to engine start (radios on):
   a. ATIS—note specified terminal conditions.
   b. Altimeter—check and set.
   c. Clock—check and set.
   d. ADF radio—tune to local station.
   e. VOR radios—check (VOT or specified VOR radial).
   f. Pre-taxi clearance—if applicable.
   g. Transponder—check and set.
   h. VOR radios—set to frequencies and courses to be employed.
   i. ELT radio—set to ARM.
3. Cockpit check after engine start:
   a. Voltmeter (or light)—check for proper operation.
   b. Ammeter—check for charge rate.
   c. Suction—check for operation.
   d. Altitude Indicator—set position of miniature aircraft.
   e. Directional Gyro—set to magnetic compass.
4. Cockpit check during taxi:
   a. Magnetic compass—check for alignment with direction of taxiway.
   b. Directinal gyro-check for alignment with mangnetic compass.
   c. ADF—check for proper station pointing.

    d. Turn Coordinator—check for proper operation (when aircraft turns.
5. Cockpit check during runup procedure:
    a. Carburetor temperature/heat—check and set.
    b. Suction—check for operation within normal limits.
    c. Directional gyro alignment—check.
    d. Altimeter—recheck.
    e. Pitot heat—on if needed.
    f. Communication radios—set to proper frequencies.
    g. Transponder—on (takeoff position).

Content—Part B.

During Part B the student shall be introduced to the basic principles of aircraft control for IFR flight.
1. Attitude indicator familiarization:
    a. Pitch indicator—demonstrate.
    b. Bank indicator (skypointer)—demonstrate.
2. Altimeter familiarization:
    a. Supporting pitch indicator—demonstrate.
    b. Lag error—demonstrate.
3. Vertical speed indicator familiarization:
    a. Trend indicator—demonstrate.
    b. Climb/descent indicator—demonstrate.
    c. Relate pitch/power settings to VSI.
    d. Relate to altimeter movement.
4. Airspeed indicator familiarization:
    a. Relate pitch to airspeed, constant power.
    b. Relate pitch to airspeed, variable power.
5. Directional gyro familiarization:
    a. Set and check procedure—demonstrate.
    b. Use as a turn indicator—demonstrate.
6. Turn coordinator familiarization:
    a. Relationship to bank—demonstrate.
    b. Slip-skid indication—demonstrate.
7. Magnetic compass:
    a. Northerly/Southerly turn error—demonstrate.
    b. Use of compass calibration card in setting directinal gyro.
8. Instrument scan—demonstrate.
9. Student practice—turns to headings as directed by the instructor.

Completion Standards. The lesson will be satisfactorily completed when the student can demonstrate the ability to make standard rate turns to specified headings within ±10° and maintain altitude within ±100 ft. Estimated instruction time: 3 hours.

Lesson No. 2 — Climbs and Descents

Objective. To use the concepts of Lesson No. 1 to control the flight of the aircraft in all three axes.

Content.

1. Review bank control, attitude control, and instrument scan techniques.
2. Climbs and descents:
   a. Maximum power climbs.
   b. Maximum power climbing turns.
   c. Reduced power descents.
   d. Reduced power turning descents.
3. Medium turns (30° bank):
   a. 360° turns.
   b. 720° turns.
4. Recoveries from unusual attitudes.
5. Introduction to partial panel flight. Demonstrate airspeed control of pitch and turn coordinator control of bank.

Completion Standards. This lesson will be satisfactorily completed when the student demonstrates the ability to maintain heading within ±10° and altitude within ±100 ft. in response to instructor assigned headings and climb/descent altitude limits or execute pattern A (Fig. 9-1). Estimated instruction time: 2 hours.

Lesson No. 3 — Precision Turns, Climbs, and Descents

Objective. To improve the student's control of the aircraft and instrument scan abilities.

Content.

1. Review scan technique and partial panel flight.
2. Timed turns—demonstrate with and without turn coordinator; turns to headings based on time only.
3. Steep turns (45° bank):
   a. 360° turns
   b. 720° turns $\Big\}$ Beginning and ending at a prescribed heading
   c. Figure eights
4. Compass turns (directional gyro covered):
   a. Turns to cardinal headings.
   b. Turns to 45° reference points.
5. Controlled descents, specified VSI values:

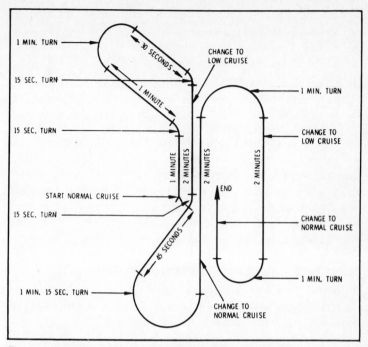

Fig. 9-1. Elements of IFR approaches and holding "in the flat."

    a. Demonstrate relationship between power and descent rate.

    b. Use of flaps to control airspeed during a descent.

  6. Primary panel. Practice turns, climbs, descents.

Completion Standards. For satisfactory completion of this lesson the student must demonstrate the ability to fly a course such as that of pattern B (Fig. 9-2) with altitude limits as prescribed by the instructor. Estimated instruction time: 1.5 hours.

Lesson No. 4 — Slow Flight

Objective. To add to the student's mastery of flight solely by reference to instruments.

Content.

  1. Review the elements of instrument scanning.

  2. Slow flight at V_{MC} + 10 knots:

    a. Half standard rate turns to headings.

    b. Timed half standard rate turns.

    c. Half standard rate compass turns.

  3. Slow flight at V_{MC}. Half standard rate turns to headings.

4. Stalls and recovery:
   a. Power off, straight ahead.
   b. Partial power, straight ahead.
   c. Partial power, climbing turning.
5. Descents at approach speeds:
   a. Partial flaps.
   b. Full flaps.
   c. Go around procedure.
6. Partial panel slow flight—review of selected maneuvers from items 1-5 preceding.

Completion Standards. This lesson will be satisfactorily completed when the student demonstrates mastery of the aircraft in slow flight by performing instructor specified maneuvers to a heading tolerance of ±10° and an altitude tolerance of ±100 ft. Estimated instruction time: 1.5 hours.

Lesson No. 5— Review of Basic Instrument Maneuvers

Objective. The objective of this lesson is to review the student's mastery of those items listed in Lessons 1 through 4 with emphasis on recovering from unusual attitudes.

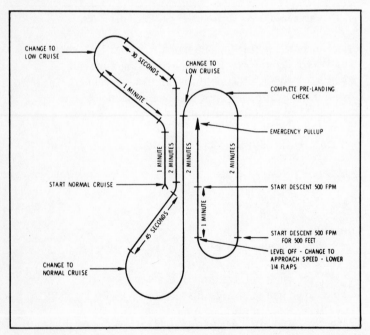

Fig. 9-2. Elements of IFR approaches and holding with climbs and descents.

Content.
1. Lesson Review.

   The instructor shall select samples of flight maneuvers learned by the student during Lessons 1 through 4 and exercise the student accordingly. Student assignments shall be given briskly to determine the student's work load limit as well as examine the student's knowledge of gaps in lesson material presented through Lesson 4.
2. Unusual Attitudes.

   The instructor shall provide the student a series of unusual attitudes from which to recover.

Completion Standards. This lesson will be satisfactorily completed when the student demonstrates mastery of the aircraft to the tolerances noted in Lessons 1 through 4. In addition, the student shall demonstrate the ability to cope with a modest amount of stress as induced by an increased work load. Estimated instruction time: 1 hour.

Lesson No. 6 —VOR Navigation

Objective. To develop the student's VOR navigational ability and also increase the scope of the student's scan.

Content.
1. Review scanning including the VOR.
2. Review precision pattern flight by flying pattern B at instructor assigned altitudes.
3. VOR course tracking:
   a. Review VOR tuning and ident. procedures.
   b. Inbound tracking to the VOR.
   c. Outbound tracking on an assigned radial.
   d. Close-in tracking techniques—VOR crossings.
   e. Intersect an assigned radial inbound and outbound.

Completion Standards. For satisfactory completion of this lesson the student must be able to hold headings within ±10°, altitudes within ±100 ft. and airspeed within ±10K. He must be able to set his course selector correctly for the course he wishes to track and to track the desired course without excessive weaving. He must demonstrate a good understanding of radial interception procedure and be able to perform simple interceptions correctly and efficiently. Estimated instruction time: 1.5 hours.

Lesson No. 7—Partial Panel Review/VOR Navigation

Objective. To further improve control of the airplane through instrument reference and scanning techniques, to develop proficiency in VOR navigation.

Content.
1. Partial panel exercise:
   a. Standard rate turns at constant altitude.
   b. Standard rate turns with climbs and descents.
   c. Double standard rate turns.
   d. Recovery from spirals.
2. VOR course tracking:
   a. Review of basic tracking techniques.
   b. Review close-in tracking techniques.
   c. Review crossing VOR and proceeding outbound on new course.
3. Radial shifts of 10°-15°, direct intercepts.
4. L shaped navigation route (intersection of two VOR radials).

Completion Standards. The proficiency level for this lesson requires headings accuracy within ±10°, altitudes within ±100 ft. and airspeed within ±10K. Also required are correct setting of the course selector, tracking which is essentially accurate, the ability to follow a route involving the intersection of courses from two VOR stations and the efficient performance of radial shifts (simple intercepts). Estimated instruction time: 1.5 hours.

Lesson No. 8 — Advanced VOR Navigation

Objective. To refine the student's VOR navigation ability and develop orientation technique.

Content.
1. Radial intercepts from unknown positions, outbound and inbound at 30° and 45° intercept angles.
2. Cross bearing fix techniques (VOR):
   a. Plotting.
   b. While tracking on a radial.
   c. With unknown position.
3. Time-distance check.

Completion Standards. Before progressing from the elements of this lesson the student must be able to fly an instructor assigned flight pattern without incurring repeated deviations greater than ±10° in heading and ±10 K in airspeed. Rate of turn must be accurate and timing within ±10 seconds. Headings flown for radial intercepts must be consistently correct. The student must display a basic understanding of time-distance checks and cross bearing fix techniques. Estimated instruction time: 1.5 hours.

Lesson No. 9—Unusual Attitudes

Objective. To develop the student's ability to cope with various types of unusual attitudes while in an IFR environment.

Content

1. Slow flight at V_{MC} with and without flaps.
2. Stall recoveries:
   a. Power off, straight ahead.
   b. Power on, straight ahead.
   c. Power on, climbing and turning.
   d. Approach.
3. Steep turns:
   a. 360° turn (45° bank).
   b. 720° turn (60° bank).
   c. Figure eights (60° bank).
4. Instructor induced unusual attitudes:
   a. Descending spiral.
   b. Steep climb/descent.
   c. Other, as needed.

Completion Standard. This lesson will be completed when the student demonstrates smooth and well controlled recoveries from the various unusual attitude situations contained in the lesson. Estimated instruction time: 1.5 hours.

Lesson No. 10—First Progress Check

Objective. To measure how well the student has mastered basic flight techniques.

Content.

This check is for the purpose of insuring that the student has mastered control of the aircraft to the degree required for progressing to Phase Two where training will be concentrated to a much greater degree on radio navigation. In order to make satisfactory progress in Phase Two, it is important for aircraft control to have been mastered to the point where great concentration is not required in merely maintaining precise heading, altitude, and airspeed control. To ensure that the above will occur, a high degree of capability in aircraft control must be demonstrated during this flight. The student will be required to demonstrate at least the following:

1. Equipment preflight check. Proper use of instruments in flight.
2. Basic aircraft control; precision pattern flying.
3. Recovery from unusual attitudes; partial panel:
   a. Approach to a stall.
   b. High speed spiral.

4. Steep turns, partial panel:
   a. 360° turn each direction with 45° bank.
   b. One minute turn each direction at double standard rate.
5. Compass turns to specified magnetic headings.
6. VOR tracking and radial intercepts.

Completion Standard, Acceptable Performance.

1. *Equipment Check.*

   Between engine start up and takeoff the student will be expected to check the functioning of all flight instruments. Knowledge of which instruments to check and how to perform the checks is required.

2. *Basic Aircraft Control.*

   Smooth and accurate control of the airplane throughout with only infrequent deviation from the following standards:
   a. Full panel: altitude during level segments; ±50 ft.; airspeed ±5K. Heading: within ±°. Climbs and descents: within ±100 FPM of target figure.
   b. Partial panel: altitude during level segments; ±65 ft.; airspeed within ±5K. Heading on completion of turns: within ±15° per 360° turn. Climbs and descents: within ±10 seconds of estimated time.

3. *Recovery from unusual attitudes using partial panel.*

   Prompt, smooth and accurate recovery. Any loss of, or error in, control which makes it necessary for the check pilot to take over will constitute a failure of the progress check.

4. *Steep Turns.*

   Full panel; smooth control with accurate turn rate and airspeed positioning. Altitude error allowance ±150 ft.

5. *Compass Turns.*

   Accurate turn rate control. Recovery within ±15° of designated heading.

6. *VOR Tracking:*
   a. Inbound and outbound tracking on assigned radials.
   b. Direct intercepts to radials (within 20° of the radial where the aircraft is located).
   c. Position determination by triangulation.
   d. Flying an L shaped intersection of 2 radials.

Estimated evaluation time: 1.5 hours.

Lesson No. 11 — Holding at a VOR

Objective. To develop proficiency in the performance of holding patterns and to develop an understanding of holding pattern entries. To develop IFR clearance comprehension and to provide training in ATC communications.

Content.

1. Use of ATC Communications (simple clearance to VOR).
2. Execution of IFR clearance (to VOR).
3. Holding over a VOR.
   a. With direct entries.
   b. With teardrop entries.
   c. With parallel entries.
   d. With strong winds.
4. Holding over an intersection:
   a. Direct entries.
   b. Teardrop entries.
   c. Parallel entries.
5. Radial interceptions.
6. Reception and execution of an alternate clearance while in flight.

Completion Standards. This lesson will be satisfactorily completed when the student demonstrates accurate interpretation and execution of a simple clearance, good progress in working out holding pattern entries and in making correct responses to high wind situations, and correct procedures for time-distance checks and radial interceptions. Estimated instruction time: 2.5 hours.

Lesson No. 12 — Simple Clearances

Objective. To provide instruction in the execution of IFR clearances involving holding patterns.

Content.

1. Simulated IFR clearance.
2. Holding over an intersection with various entries.
3. Orientation exercise:
   a. Two station fix.
   b. Time distance check.
   c. Radial intercept.
4. Holding over a VOR with various entries.
5. Execute simulated IFR clearance for return to airport.

Completion Standards. To successfully complete this lesson the student must demonstrate progress in interpretation and execution of IFR clearances, relatively accurate holding pattern entries, and precise execution of the holding pattern itself. Estimated instruction time: 1.5 hours.

Lesson No. 13 — VOR Approaches

Objective. To introduce the instrument approach concept and to develop a basic ability to perform VOR based instrument approaches.

Content.
1. Simulated ATC clearance.
2. Execute simulated IFR clearance to an airport via a VOR which will be used for an approach.
3. Holding pattern over the VOR.
4. Execution of VOR instrument approaches:
   a. VOR at airport site.
   b. VOR off airport site.
   c. Simulated radar vectors for a "straight in" VOR approach.
5. Missed approach.
6. Execute simulated IFR clearance to return to airport.

Completion Standard. This lesson will be satisfactorily completed when the student demonstrates the ability to fly VOR approaches using the correct heading and maintaining all altitude restrictions. Estimated instruction time: 3 hours.

Lesson No. 14 — Introduction to ADF

Objective. To introduce the student to ADF navigation and develop a sense of similarity with VOR navigation.

Content.
1. Simulated IFR clearance with at least one leg to be navigated by ADF tracking with a VOR instrument approach at the destination.
2. Execute simulated clearance.
3. Perform a series of VOR approaches and related missed approaches.
4. ADF tracking to and from the ADF station using instructor assigned "radials."
5. Review recoveries from stalls and unusual attitudes using partial panel.
6. Execute simulated IFR clearance to return to airport.

Completion Standard. This lesson will be satisfactorily completed when the student demonstrates the ability to fly a VOR approach within reasonable tolerances, the ability to track an ADF course, ad the ability to recover from unusual attitudes. Estimated instruction time: 2 hours.

Lesson No. 15 — NDB Holding

Objective. To further develop ADF tracking technique and to introduce NDB holding procedures.

Content.

1. Simulated IFR clearance to an NDB.
2. Execute simulated IFR clearance.
3. Track inbound "to," then outbound "from" an NDB, with large course changes over the station.
4. Intercept ADF courses, both "to" and "from," while on assigned headings.
5. Intercept ADF courses, "to" and "from," while tracking on VOR courses.
6. Perform holding practice at an NDB, various entries.
7. Execute simulated IFR clearance to return to airport.

Completion Standard. This lesson will be satisfactorily completed when the student demonstrates correct performance in ADF tracking, intercept, and holding procedures. Estimated instruction time: 2 hours.

Lesson No. 16 — ADF Approaches

Objective. To develop proficiency in ADF navigation, including the NDB based instrument approach. To obtain additional experience in performing VOR approaches.

Content.

1. Execute a simulated IFR clearance involving interception of an ADF leg and a holding assignment over a non-directional beacon.
2. Practice varied holding pattern entries at an NDB.
3. Execute ADF approaches, both "full" approaches and "straight-in" approaches, including at least one which involves simulated radar vectors to final.
4. Review VOR approaches as they relate to ADF approaches.
5. Execute simulated IFR clearance to airport using ADF navigation.

Completion Standard. To satisfactorily complete this lesson the student shall display a high degree of proficiency in ADF navigation and show good progress in performing the ADF approach. Estimated instruction time: 3 hours.

Lesson No. 17 — DME Approaches

Objective. To acquaint the student with DME arcs and VOR/DME approaches. To continue progress in ADF approaches.

Content.

1. Execute a simulated IFR clearance involving a DME arc.
2. Practice a series of DME arcs and DME/VOR approaches including approaches with a series of DME step down fixes.
3. Review ADF approaches including missed approaches.
4. Execute a simulated IFR clearance to the airport using VOR and ADF navigation.

Completion Standard. This lesson shall be completed when the student exhibits the ability to maintain a DME arc within ±0.5 miles and can successfully execute VOR/DME approaches. Estimated instruction time: 1.5 hours.

Lesson No. 18—ILS Approaches

Objective. To introduce the ILS approach and to continue progress in ADF approaches.

Content.

1. Execute simulated IFE clearance involving a course to and a holding assignment over an NDB.
2. Obtain a clearance from ATC and perform a series of ILS localizer approaches.
3. Obtain a clearance from ATC and perform a series of full ILS approaches (glide slope and localizer) from appropriate holding positions.
4. Review ADF approaches.
5. Execute an actual ATC practice clearance to the home airport.

Completion Standards. Successful completion of this lesson requires that the student exhibit a basic grasp of ILS and localizer approaches and a high level of performance in ADF approaches. Estimated instruction time: 3.5 hours.

Lesson No. 19—ILS Back Course Approaches

Objective. To substantially master the ILS approach and to develop proficiency in performing ILS back course localizer approaches.

Content.

1. Execute a simulated IFR clearance to a holding fix from which a practice ILS back course clearance may be obtained.
2. Execute the simulated clearance; obtain an ATC clearance for an ILS back course approach.
3. Perform a series of ILS back course approaches beginning at appropriate holding fixes.

4. Facilities permitting, perform one or more practice DME arc to ILS type approaches.
5. Execute an actual ATC practice clearance to return to the home airport.

Completion Standards. This lesson shall be successfully completed when the student demonstrates the ability to fly an ILS back course approach within appropriate tolerances. In addition, the student must show a complete understanding of reverse sensing. Estimated instruction time: 2.5 hours.

Lesson No. 20—Complex Clearances

Objective. To gain proficiency in the interpretation of more complicated clearances and in the taking and executing clearances received while airborne. In addition, to develop additional proficiency at making ILS or localizer approaches.

Content.
1. Interpretation and execution of a relatively complicated clearance. This clearance should include various types of courses (VOR, ADF, ILS) with at least two short legs.
2. En route clearance amendments, or changes, holding problems.
3. Front and back course ILS approaches.
4. Partial panel review:
    a. Steep turns.
    b. Stalls with 20° flap settings and high power.
    c. Unusual attitude recoveries.
5. Execution of a relatively complicated clearance to return to the home airport.

Completion Standards. To satisfactorily complete this lesson the student must demonstrate accurate execution of the clearance and clearance amendments, correct holding entries and precise holding patterns, and accurate execution of ILS approaches. Estimated instruction time: 1.5 hours.

Lesson No. 21—Radar Approaches

Objective. To acquaint the student with the radar ASR approach and, facilities permitting, the PAR approach.

Content.
1. Execute a reasonably complex clearance which involves an area arrival and a radar approach.
2. From a suitable holding point obtain an ATC clearance and perform several ASR approaches.

3. Facilities permitting, repeat the ASR appraoch as a no-gyro ASR approach.
4. Facilities permitting, perform a short series of PAR approaches.
5. Execute a reasonably complex missed approach clearance to the home airport.

Completion Standards. This lesson is complete when the student demonstrates the ability to accurately fly a radar approach and communicate with ATC in a meaningful fashion. Estimated instruction time: 1.5 hours.

Lesson No. 22—General Radio Navigation Review

Objective. To bring all radio navigation procedures thus far introduced to a high level of proficiency.

Content.
1. Execute a clearance which involves the use of all the types of facilities and all of the procedures thus far practiced. The clearances should include an area departure, an area arrival, a holding problem, en route amended clearances, and an approach.
2. Practice a time-distance problem.
3. Fix orientation using two VOR's or one VOR and one NDB.
4. Execute practice instrument approaches, using only one radio:
   a. VOR
   b. ADF
   c. ILS
   d. Localizer
   e. Back course ILS

Completion Standards. To satisfactorily complete this lesson the student must demonstrate performance of all maneuvers at Instrument Rating Flight test standards. Estimated instruction time: 2 hours.

Lesson No. 23—Second Progress Check

Objective. To measure how well the student has mastered instrument navigation techniques.

Content.
This progress check is for the purpose of insuring that the student has reached the level of proficiency necessary for advancing to the instrument cross-country phase of the course.

For this check, the student will be required to demonstrate the following:

1. Equipment and radio check.
   a. Demonstrate checking flight instruments and radio equipment for proper functioning and accuracy.
   b. Tune and index radios and instruments, as appropriate, for the departure from the airport.
2. Copying and interpreting an IFR clearance.
3. Performance of an area departure in accordance with a clearance.
4. Performance of a holding pattern entry.
5. Performance of a radial intercept.
6. Performance of an area arrival and instrument approach.
7. Two additional instrument approaches, so that in all an ILS (or backcourse LOC), plus a VOR and an ADF approach are accomplished.

Completion Standard, Acceptable Performance.

1. Performance of an accurate VOT check of VOR receivers. Correct tuning and indexing of all navigational equipment after receiving a clearance. Correct checking, while taxiing, of the functioning of the instruments noted in Lesson No. 1.
2. Clearance copying and interpretation will be evaluated on the basis of accuracy and efficiency.
3. Performance of navigational exercises will be evaluated for accuracy of track, identification of courses, and operation of the navigational equipment. Altitude maintenance for cruising operations must be substantially held within ±100 ft.
4. Holding pattern entries must be efficient and the pattern must be performed in the assigned position with proper corrections for the effects of wind.
5. The accurate performance of instrument approaches down to authorized minimums. Errors no greater than 100 ft. below prescribed altitudes during initial approach, with *no descent below minimums after passing the final fix* will be required for acceptable performance.
6. A high degree of knowledge of ATC procedures must be demonstrated. Performance of radio communications will be evaluated on the basis of the use of corrected radio phraseology, selection of the appropriate frequencies, accuracy and clarity of transmissions, and the timing of communications.
7. Evaluation of the missed approach will be based on accuracy of the procedure, timing of the decision to execute a missed approach, and the appropriateness of communications.

Estimated instruction time: 1.5 hours.

Lesson No. 24 — Transition to a High-Performance Aircraft

Note: In the event a student has performed preliminary work in a high-performance aircraft, this lesson is not necessary.

Objective. To develop proficiency in performing instrument navigation procedures in a high performance aircraft. This lesson is to be preceeded by a thorough briefing on the aircraft and its equipment.

Study Reference: Owners Manual for the aircraft to be used in future IFR navigational training.

Content.
1. Check equipment, tune radios.
2. Instrument flight familiarization:
   a. Normal climb.
   b. Level cruise, normal speed.
   c. Steep turns.
   d. Slow cruise.
   e. Slow speed descents.
   f. Normal cruise speed descents.
   g. Stalls with and without flaps.
3. Familiarization, primary panel instruments:
   a. Straight and level.
   b. Standard rate turns.
   c. Steep turns.
   d. Climbs.
   e. Slow speed descents.
   f. Cruise speed descents.
   g. Stalls with and without flaps.
   h. Unusual attitude recoveries.
4. Fly pattern "B."
5. Execute an instrument clearance involving an area departure, amended clearance problem, and a holding assignment.
6. Navigational radio emergencies.
7. Perform area arrivals, terminating in instrument approaches, and missed approaches (DME arcs incorporated where practical).

Completion Standards. To satisfactorily complete this lesson all maneuvers must be performed by the student to the same standards required in the second progress check (Lesson 23). Estimated instruction time: 3 hours.

Lesson No. 25—First Practice Cross-Country Flight

Objective. To develop proficiency in planning and executing IFR cross-country flights.

Content.

1. Plan an IFR flight between two airports with IFR approaches (approximately 100 NM apart). Develop written flight logs for the outbound and return routes.
2. Make out and file an instrument flight plan for the flight (each direction).
3. Perform the IFR flight which has been planned, including an instrument approach at each destination. The entire flight must be accomplished in conformance with an ATC clearance and normal ATC communications.

Completion Standards. The student shall demonstrate good progress in the performance of the IFR cross-country to satisfactorily complete this lesson. Estimated instruction time: 2 hours.

Lesson No. 26—Second Practice Cross-Country Flight

Objective. To continue building instrument cross-country experience and proficiency in preparation for advanced instrument navigation work.

Content.

1. Plan an IFR flight between two major airports with IFR approaches (approximately 100 NM or more apart). Develop written flight logs accordingly.
2. Make out and file an instrument flight plan for the flight (each direction).
3. Perform the IFR flight which has been planned, including an instrument approach at each destination. The entire flight must be accomplished in accordance with an ATC clearance and normal ATC communications.

Completion Standards. Demonstration of sufficient proficiency in instrument cross-country flying that a maximum of value will be gained from advanced cross-country flights. Estimated instruction time: 3 hours.

Lesson No. 27—Advanced Cross-Country Flight

Objective. To develop the student's level of proficiency and experience in operating an aircraft on instruments under instrument flight rules in a high density terminal environment to an acceptable level. Whenever practicable the student will, in this phase, operate the aircraft in actual instrument conditions.

Content.

1. Plan an extended cross-country flight with landings at no less than three different airports with IFR approaches, preferably in a high density area. Plan to perform no less than one of each of the following types of approaches:

   a. VOR.

   b. VOR/DME.

   c. NDB.

   d. ILS localizer.

   e. ILS.

   f. ILS back course.

   The cross-country must be at least 250 nautical miles in distance per FAR. A distance of at least 500 nautical miles is suggested.

2. Make out and file flight plans for all routes to be utilized, including the naming of an alternate airport.

3. Perform the IFR flights which have been planned. All segments must be accomplished in conformance with ATC clearances, using normal ATC communications.

Completion Standards. Prior to the termination of this trip the student will be expected to demonstrate the ability to plan, file for, and conduct an IFR flight (including copying and interpreting clearances) with a high level of accuracy and without assistance from the instructor.

In conducting the flight, the student must be able to properly check, set and tune all equipment, perform the departure, en route, and arrival phases, including instrument approaches, in a knowledgable and correct manner. Accuracy of course and altitude holding must be at a level compatible with instrument flight test standards. Estimated instruction time: 5.5 hours.

Lesson No. 28 — Final IFR Review

Objective. To correct any deficiencies which may exist in the student's proficiency. To ascertain that he has reached a graduation level.

Content. Review and practice all maneuvers and procedures for which such work is required.

Completion Standards. To satisfactorily complete this lesson, the student must perform all maneuvers and procedures at final progress check standards or better. Estimated instruction time: 1.5 hours.

Lesson No. 29 — Final IFR Progress Check

Objective. This progress check is for the purpose of insuring that the student has reached IFR course graduation standards.

Content.

For this check the student must demonstrate the following:

1. Equipment and radio check procedures.
2. Radio navigation:
   a. Receive and interpret a clearance involving at least four courses, one of which is an ADF course.
   b. Execute the above clearance.
   c. Execute a holding pattern entry and fly a holding pattern.
   d. Perform an orientation by fix or time-distance check method. (Exclude DME and radar).
   e. Perform radial intercepts.
   f. Demonstrate tracking of a DME arc.
   g. Execute an area arrival and an instrument approach.
   h. Execute a missed approach.
   i. Execute two additional approaches. (Include ILS (or BC LOC), ADF or VOR).
   j. Use ATC (communications throughout the exercise.
   k. Navigational radio malfunctions.
3. Aircraft control:
   a. Straight and level.
   b. Climbs and descents.
   c. Maneuvering at approach speed.
   d. Steep turns (360° each direction with 45° bank).
4. Partial panel aircraft control:
   a. Steep turns (approximately double standard rate).
   b. Compass turns.
   c. Stall recoveries.
   d. Recoveries from unusual attitudes.

Completion Standards, Acceptable Performance. Performance of all maneuvers and procedures must be on a level at least equal to that required in the Instrument Pilot Flight Test Guide. In addition, the student must demonstrate knowledge of the static system and transponder checks required by FAR and determine that the aircraft used for the flight check has the required logbook entries. The proper logbook entry must also be demonstrated for a VOR accuracy check. Estimated instruction time: 1.5 hours.

Note: The following series of seven lessons are devoted to maneuvers required for the Commercial Certificate. In general, a flight school will integrate these maneuvers in with the IFR maneuvers when a candidate is working toward his Commercial Certificate. A typical integration plan is provided at the beginning of this chapter.

Lesson No. 30—Review of Private Pilot Maneuvers

Objective. Through dual instruction and solo practice, the objective of this lesson is to review and bring to a skilled performance level those VFR maneuvers required for the Private Pilot Certificate.

Content.
1. Aircraft preflight inspection/engine runup.
2. Basic aircraft control:
    a. Flight at minimum controllable airspeed.
    b. Power on/off stalls.
    c. Climbing-turning stalls.
    d. Approach stalls and go-arounds.
3. Landings and takeoffs:
    a. Normal.
    b. Short field.
    c. Soft field.
    d. Crosswind.
    e. Simulated flat failures.
    f. Emergency landings.
    g. Slips to a landing.
4. Maneuvers:
    a. Steep turns (45° bank).
    b. Turns about a point.
    c. S turns across a road.
    d. Spirals about a point.
5. Advanced maneuvers:
    a. Accelerated stalls.
    b. Recovery from unusual attitudes.
    c. Pylon eights (optional).
    d. Spin recovery (optional).

Completion Standards. The successful completion of this lesson will be demonstrated by student performance of all maneuvers listed at or above the level required for the Private Pilot Certificate. Estimated instruction time: 3 hours. Estimated solo practice time: 3 hours.

Lesson No. 31—Commercial Pilot Flight Maneuvers

Objective. To introduce and develop proficiency in those advanced flight maneuvers required for the Commercial Certificate. This lesson includes both dual instruction and solo practice.

Content.

1. Preflight of a complex aircraft.
2. Steep turns (60° bank angles):
    a. 360° turns.
    b. 720° turns.
    c. Figure eights.
3. Chandelles.
4. Lazy eights:
    a. 15° bank angle maximum.
    b. 30° bank angle maximum.
    c. 45° bank angle maximum.
5. Slow flight at V_{MC} with various flap settings.
6. Accuracy landings:
    a. With power.
    b. Without power.
7. Steep spirals about a point.
8. Recovery from unusual attitudes.
9. Emergency procedures relative to a complex aircraft:
    a. Simulated landing gear failure.
    b. Simulated flap failure.
    c. Simulated power failure.
10. Related complex aircraft operations:
    a. Turbosupercharger operation precautions during landings and takeoffs.
    b. Oxygen system operation.
    c. Autopilot operation.

Completion Standards. The successful completion of this lesson will be demonstrated by student performance of all maneuvers and operations listed at the level of performance required for the Commercial Certificate. Estimated instruction time: 4 hours. Estimated solo practice time: 4 hours.

Lesson No. 32 — Night Flight

Objective. To provide additional experience, confidence, and skill in night flying techniques.

Content.

1. Night flight preparation, including aircraft and equipment check.
2. Takeoff and landing practice (full stop landings) at a minimum of three airports (as assigned by instructor) in addition to home airport.
3. Preplanning and practice navigation to at least one of the airports of intended landing by tracking VOR radials.

Completion Standards. This lesson is complete when the student has satisfactorily accomplished the specified practice. Estimated instruction time: 1 hour. Estimated solo practice time: 2 hours.

Lesson No. 33 — VFR Cross-Country Flight

Objective. To develop the student's proficiency in VFR cross-country flight.

Content.

1. Plan a cross-country flight which will furnish appropriate VFR training and flight experience:
   a. Obtain airport, navigational aid, NOTAM, and weather information from appropriate sources.
   b. Prepare a VFR navigation planning log.
   c. Review flight planning with instructor.
   d. File flight plan.
2. Conduct the planned cross-country flight using dead reckoning, pilotage, and radio navigation.
3. Update and complete navigation log throughout the flight.
4. Compute actual fuel consumption rates and reserves. Review with flight instructor at the end of the cross-country.

Completion Standards. This lesson is satisfactorily complete when the student has accomplished the assigned cross-country navigation flight and reviewed the navigation log and fuel consumption figures with the flight instructor. Estimated dual instruction time: 3 hours. Estimated solo practice time: 3 hours.

Lesson No. 34 — Extended VFR Cross-Country Flight

Objective. To perform an expanded cross-country navigation flight providing additional skill improvement and judgment development through a practical exercise.

Content.

1. This lesson will consist of cross-country navigation routes with landings at three airports, each of which is approximately 200 NM from each of the other two. Conditions permitting, at least one airport should be serviced with a TRSA or TCA.

2. Preplanning will include:
   a. Selection and plotting of routes.
   b. Completion of navigation log for each route, including estimate of fuel remaining after each flight.
   c. Comprehensive check of current and forecast weather conditions.
   d. Check of appropriate sources to develop adequate information concerning each airport and each radio aid to be used.

3. Conduct the planned flight using dead reckoning, radio navigation, and pilotage.

4. Post flight review with instructor.

Completion Standards. This lesson is satisfactorily complete when the student has performed the assigned flight and the flight has been reviewed with the instructor. Estimated solo practice time: 5 hours.

Lesson No. 35—Night VFR Cross-Country Flight

Objective. To provide night solo cross-country flight experience during which the student will use knowledge and skills acquired in VFR and IFR training.

Content.

1. Plan a night cross-country navigation flight to an airport at least 100 NM away and a return flight:
   a. Obtain airport, navigational aid, NOTAM, and weather information for the flight.
   b. Prepare VFR navigation planning log.
   c. Review flight planning with instructor.
   d. Perform night flight equipment check.

2. Conduct the planned cross-country flight using pilotage, dead reckoning, and radio navigation.

3. Update and complete navigation log throughout the flight.

4. Compute actual fuel consumption rates and reserves.

5. Post flight review with instructor.

Completion Standards. This flight is satisfactorily complete when the student has accomplished the assigned night cross-country navigation flight and reviewed the navigation log with the flight instructor. Estimated solo practice time: 3 hours.

Lesson No. 36 — Final Commercial VFR Progress Check

Objective. This progress check is for the purpose of insuring that the student has reached the Commercial Pilot Certificate standards in so far as VFR flight is concerned. The student shall demonstrate that he is capable of cross country-flight as well as commercial flight maneuvers.

Content—Part A: Cross-Country Flight Planning.

1. Plan a cross-country flight per instructor assigned routing and related load (passenger) assignments.
2. Prepare a flight plan.
3. Perform a weather check.
4. Check all required aircraft records.
5. Demonstrate cross-country navigation involving the use of radio aids, dead reckoning, and pilotage.
6. Per instructor assignment, change routing to an alternate destination; include revised ETA and revised flight plan.

Note: Generally, the cross-country flight demonstration consists basically of exiting the airport and flying a portion of the initial leg of the flight.

Content—Part B: Commercial Maneuvers.

1. Slow flight at V_{MC} with varying flap settings.
2. Stall series:
   a. Departure stalls (imminent).
   b. Approach stalls (imminent and full).
   c. Accelerated (imminent).
3. Steep turns (60° bank angle).
4. Chandelles.
5. Lazy eights.
6. Steep spiral about a point.
7. On-pylon eights (optional).
8. Spin recovery (optional).
9. Short field approaches, with:
   a. Power off landings.
   b. Power on (rough field) landings.
   c. Emergency go-arounds.
10. Takeoffs:
    a. Normal.
    b. Short field: obstacle.
    c. Soft field.
11. Precision landings (within 200 ft. of an assigned point).
12. Slips to landings.

Completion Standards, Acceptable Performance. For successful completion of this phase test, the student must demonstrate knowledgeable and accurate performance in navigation. For landing and takeoff demonstrations, safe and correct procedures performed with a high degree of accuracy are required. Estimated instruction time: 1.5 hours.

Concluding Note

Once again it is stressed that the preceeding series of 36 "lessons" are simply blocks of learning and do not constitute actual lesson plans. The lesson plan is an individual event, planned *specifically for you*, by your flight instructor. It is based upon a syllabus such as the preceding but is unique to *your* needs, *your* airport, *your* training aircraft, and *your* local flying conditions. In this regard, special blocks of learning and lesson plans should be prepared for situations such as the following:

1. Mountain flying:
   a. Density altitude, high altitude airport landing and takeoff operations.
   b. Mountain wind current.
   c. Entrapment in canyons, flying mountain passes.
   d. Route selection, avoidance of wilderness areas.
2. Cold weather operations:
   a. Engine preheat, airframe defrost.
   b. Aircraft icing, freezing rain, snow.
3. Tropical flight:
   a. Thunderstorm avoidance.
   b. Fog, heavy rain.
   c. Route planning, avoidance of swamp areas.
4. Desert flight:
   a. Hi-temperature engine operation.
   b. Route planning, avoidance of wilderness areas, emergency water.
5. High-traffic density areas:
   a. Tower-to-tower IFR clearances.
   b. Preferred routes.
   c. Local area radio procedures.
6. Seacoast/Lake areas:
   a. Saltwater corrosion protection.
   b. Over water flight precautions, FAA radio services available.
   c. Thunderstorm avoidance, fog, heavy rain.
   d. Ditching procedures.
   e. Emergency flotation equipment.

7. Operations involving unattended airports:
   a. Reserve fuel allowance.
   b. Carriage of spare oil, emergency tools, spare alternator belt, etc.
8. Flight to Mexico or Canada:
   a. Fuel contamination problems, precautionary procedures (Mexico).
   b. Border crossing flight plan requirements.
   c. Customs procedures.
   d. Mexican/Canadian flight rules.
   e. Mexican/Canadian insurance requirements.
9. Other flying conditions:
   a. Airports with excessive landing fees.
   b. Landing off airports, legal considerations.
   c. Landing on unimproved runways.
   d. Appropriate emergency supplies.

PART 3: THE INSTRUMENT FLIGHT TEST

(Extracted from the *Instrument Pilot-Airplane Flight Test Guide* AC 61-56A.)

Use of the Flight Test Guide

The pilot operations in this flight test guide, indicated by Roman numerals, are required by § 61.65 of Part 61. This guide is intended only to outline appropriate pilot operations and the minimum standards for the performance of each procedure or maneuver which will be accepted by the examiner as evidence of the pilot's competency. With the exception of Pilot Operation III (Instrument Approaches. FAR 61.65(c) (3)), it is not intended that the applicant be tested on every procedure or maneuver within each pilot operation, but only those considered necessary by the examiner to determine competency in each pilot operation. Throughout the flight test, certain procedures or manuevers may be evaluated separately or in combination with other procedures or maneuvers.

This guide contains an *Objective* for each required pilot operation. Under each pilot operation, pertinent procedures or maneuvers are listed with *Descriptions* and *Acceptable Performance Guidelines*.

1. The *Objective* states briefly the purpose of each pilot operation required on the flight test.

2. The *Description* provides information on what may be asked of the applicant regarding the selected procedure or maneuver. The procedures or maneuvers listed have been found most effective in demonstrating the objective of that particular pilot operation.

3. The *Acceptable Performance Guidelines* include the factors which will be taken into account by the examiner in deciding whether the applicant has met the objective of the pilot operation. The airspeed, altitude, and heading tolerances given represent the minimum performance expected in good flying conditions. However, consistently exceeding these tolerances before corrective action is initiated or prematurely descending below DH or MDA, is indicative of an unsatisfactory performance. Any procedure or action, or the lack thereof, which requires the intervention of the examiner to maintain safe flight will be disqualifying.

In the event the applicant takes the instrument pilot flight test and the commercial pilot flight test simultaneously, the maneuvers selected by the examiner for each may be combined and evaluated together, where practicable.

General Procedures For Flight Tests

The ability of an applicant for an instrument pilot airplane rating to perform the required pilot operations is based on the following:

1. Completing a checklist for instrument flight operations appropriate to the airplane and equipment used.

2. Performing procedures and maneuvers within the airplane's performance capabilities and limitations, including use of the airplane's systems.

3. Performing emergency procedures and maneuvers appropriate to the airplane used.

4. Piloting the airplane with smoothness and accuracy.

5. Exercising judgment.

6. Applying aeronautical knowledge.

7. Displaying mastery of the airplane, with the successful outcome of a procedure or maneuver never seriously in doubt.

Failure of any required pilot operation is a failure of the flight test. The examiner or the applicant may discontinue the test at any time when the failure of a required pilot operation makes the applicant ineligible for the certificate or rating sought. If the test is discontinued, the applicant is entitled to credit for only those entire pilot operations that were successfully performed.

Flight Test Prerequisites

An applicant for the instrument pilot airplane flight test is required by § 61.39 of the Federal Aviation Regulations to have:

1. Passed the Instrument Pilot Airplane Written Test within 24 months before the date of the flight test.

2. The applicable instruction and aeronautical experience prescribed in Part 61.

3. At least a third class medical certificate issued within the past 24 months.

4. A written statement from a certificated instrument flight instructor certifying that:

    a. The applicant has been given flight instruction in preparation for the flight test within 60 days preceding the date of application.

    b. The applicant is found competent to pass the flight test and to have a satisfactory knowledge of the subject areas in which deficiency is shown by the airmen written test report.

Airplane and Equipment Requirements for Flight Test

The applicant is required by FAR 61.45 to provide an air worthy airplane for the flight test. This airplane must be capable of, and its operating limitations must not prohibit, the pilot operations required on the flight test. Flight instruments required are those appropriate for controlling the airplane without outside references. The required radio equipment is that necessary for communications with ATC and for the performance of VOR, ADF, and ILS (glide slope and localizer) approaches.

I. Maneuvering by Reference to Instruments

Objective: To determine that the applicant can safely and accurately maneuver the airplane in instrument conditions.

A. Straight-and-Level Flight.

1. Description. The applicant may be asked to demonstrate straight-and-level flight with changes in airspeed and airplane configuration. The applicant will be expected to maintain altitude and heading and to accurately control airspeed.

2. Acceptable Performance Guidelines. The applicant's performance shall be evaluated on the ability to maintain altitude within ±100 ft., heading within ±10°, and airspeed within ±10 kts. of that assigned.

B. Turns.

1. Description. The applicant may be asked to demonstrate heading changes using various means to determine rate and amount of turn in level, climbing, and descending flight. This may include changes in airspeed and airplane configuration. Turns for this demonstration may be selected from the following:

    a. Standard rate turns.

    b. Timed turns.

    c. Turns to predetermined headings.

    d. Magnetic compass turns.

    e. Steep turns.

2. Acceptable Performance Guidelines. The applicant's performance shall be evaluated on the ability to complete level, climbing, and descending turns to within ±10° of the desired heading while maintaining the desired airspeed within ±10 kts. In level turns, the applicant shall maintain desired altitude within ±100 ft. Climbing and descending turns shall be completed within ±100 ft. of the desired altitude. Changes of airspeed and airplane configuration shall be completed within ±100 ft. of the desired altitude and ±10 kts. of the desired airspeed.

C. Climbs and Descents.

1. Description. The applicant may be asked to demonstrate changes of altitude including:

    a. Constant airspeed climbs and descents.

    b. Rate climbs and descents.

    c. Climbs and descents to predetermined altitudes and headings.

The examiner may request that the above demonstrations be performed in various airplane configurations.

2. Acceptable Performance Guidelines. The applicant's performance shall be evaluated on the ability to maintain airspeed within ±10 kts. and vertical rate within ±200 ft. per minute of that desired. Leveloffs and rollouts shall be completed within ±100 ft. and ±10° of the altitude and heading assigned.

II. IFR Navigation

Objective: To determine that the applicant can safely and efficiently navigate in instrument conditions in the National Airspace System in compliance with Instrument Flight Rules and ATC clearances and instructions.

A. Time, Speed, and Distance.

1. Description. The applicant may be asked to demonstrate preflight and inflight computations to determine ETE, ETA, wind correction angle, and groundspeed.

2. Acceptable Performance Guidelines. The applicant's performance shall be evaluated on the basis of the ability to make accurate and timely computations.

B. VOR Navigation.

1. Description. The applicant may be asked to demonstrate:

a. Intercepting a VOR radial at a predetermined angle.

b. Tracking on a selected VOR radial.

c. Determining position using intersecting VOR radials.

2. Acceptable Performance Guidelines. The applicant's performance shall be evaluated on the basis of accuracy in determining position by means of cross bearings, interception procedures, and ability to maintain orientation and the assigned flight path.

C. ADF Navigation.

1. Description. The applicant may be asked to use ADF for homing, intercepting, and tracking predetermined radio bearings to and from non-directional beacons, and for determining position by use of cross bearings.

2. Acceptable Performance Guidelines. The applicant's performance shall be evaluated on the basis of accuracy in determining position by means of cross bearings, interception procedures, and the ability to maintain orientation and the assigned track.

D. Navigation by ATC Instructions.

1. Description. The applicant may be asked to demonstrate the ability to comply with ATC instructions and procedures. This includes navigation by adherence to radar vectors and specific instructions for headings and altitude changes.

2. Acceptable Performance Guidelines. Evaluation of the applicants performance shall be based on the promptness and accuracy shown in responding to and complying with ATC navigation instructions.

III. Instrument Approaches

Objective: To determine that the applicant can execute safe and accurate instrument approaches to published minimums under instrument conditions.

A. VOR Approach.

1. Description. The applicant will be required to demonstrate a published VOR approach procedure.

2. Acceptable Performance Guidelines. The applicant shall descend on a course so as to arrive at the MDA at or before the missed approach point, in a position from which a normal landing approach can be made, straight-in or circling, as appropriate. The missed approach point shall be determined by accurate timing from the final approach fix. Deviations of more than ±10 kts. from the desired approach speed shall be disqualifying. Descent below minimum altitudes during any part of the approach or descent below the MDA prior to the examiner reporting the runway environment in sight, shall be disqualifying. If a circling approach is made, exceeding the radius of turn dictated by published visibility minimums or descending below the MDA prior to reaching a position from which a normal approach to the landing runway can be made, shall also be disqualifying.

B. ILS Approach.

1. Description. The applicant will be required to demonstrate a published ILS approach procedure.

2. Acceptable Performance Guidelines. As directed by the examiner, the applicant shall descend on a straight-in approach to the DH, or on a circling approach to the MDA, arriving in a position from which a normal landing approach can be made straight-in or circling, as appropriate. Deviations of more than ±10 kts. from the desired approach speed shall be disqualifying. Descent below minimum altitudes during any part of the approach, full scale deflection of the CDI or the glide slope indicator after glide slope interception or descent below the DH or MDA prior to the examiner reporting the runway environment in sight, shall be disqualifying. If a circling approach is made, exceeding the radius of turn dictated by published visibility minimums or descending below the MDA prior to reaching a position from which a normal approach to the landing runway can be made, shall also be disqualifying.

C. ADF Approach.

1. Description. The applicant will be required to demonstrate an ADF approach using a published NDB (non-directional beacon) approach procedure.

2. Acceptable Performance Guidelines. The applicant shall descend on a course so as to arrive at the MDA at or before the missed approach point, in a position from which a normal landing approach can be made, straight-in or circling, as appropriate. The missed approach point shall be determined by accurate timing from the final approach fix. Deviations of more than ±10 kts. from the desired approach speed shall be disqualifying. Descent below minimum altitudes during any part of the approach or descent below the MDA prior to the examiner reporting the runway environment in sight, shall be disqualifying. If a circling approach is made, exceeding the radius of turn dictated by published visibility minimums or descending below the MDA prior to reaching a position from which a normal approach to the landing runway can be made, shall also be disqualifying.

IV. Cross-Country Flying[1]

Objective: To determine that the applicant can competently conduct en route and terminal operations within the National Airspace System in instrument conditions, using radio aids and complying with ATC instructions.

A. Selection of Route.

1. Description. The applicant may be asked to select a route for a 250 nautical mile IFR flight, based on information contained in the *Airman's Information Manual, En route Charts, Instrument Approach Procedure Charts,* and other appropriate sources of information. This includes facilities for all departures and arrivals.

2. Acceptable Performance Guidelines. The applicant's performance shall be evaluated on the ability to obtain and apply pertinent information for the selection of a suitable route. Failure to determine current status and usability of facilities shall be disqualifying.

B. Procurement and Analysis of Weather Information.

1. Description. The applicant may be asked to procure and analyze weather reports and forecasts pertinent to the proposed flight. This information should provide (1) forecast weather conditions at destination, (2) the basis for selecting an alternate airport, and (3) the basis for selecting a route to avoid severe weather.

2. Acceptable Performance Guidelines. The applicant shall correctly analyze the weather reports and forecasts and understand their significance to the proposed flight. Failure to recognize conditions which would be hazardous to the flight shall be disqualifying.

C. Development of Flight Log.

1. Description. The applicant may be asked to develop a flight

---

[1]The examiner may ask the applicant to plan an IFR cross-country flight and set out on course. The flight may be continued only long enough for the examiner to determine the applicant's competence in IFR cross-country flying.

log for the proposed flight. This log should include at least the en route courses, estimated ground speeds, distances between checkpoints, estimated time between checkpoints (ETEs), and amount of fuel required. On the basis of the log, the applicant is expected to prepare an IFR flight plan for the examiner's review.

2. Acceptable Performance Guidelines. The applicant's performance shall be evaluated on the completeness and accuracy of the flight log and flight plan.

D. Aircraft Performance and Limitations.

1. Description. The applicant may be asked to apply the information contained in the airplane flight manual or manufacturer's published recommendations to determine the aircraft performance capabilities and weight and balance limitations.

2. Acceptable Performance Guidelines. The applicant's performance shall be evaluated on the proper application of aircraft performance and loading data in the conduct of the proposed flight.

E. Aircraft Systems and Equipment.

1. Description. The applicant may be asked to explain the use of the instruments, avionic equipment, and any special system installed in the airplane used, including indications of malfunctions and limitations of these units.

2. Acceptable Performance Guidelines. The applicant's performance shall be evaluated on knowledge of the instruments and equipment which are installed in the airplane used for the flight test.

F. Preflight Check of Instruments and Equipment.

1. Description. Prior to takeoff, the applicant may be asked to perform a systematic operational check of engine instruments, flight instruments, and avionic equipment. All equipment should be appropriately set for the departure clearance requirement.

2. Acceptable Performance Guidelines. The applicant's performance shall be evaluated on the thoroughness and accuracy of the checks and procedures. Failure to properly check and set instruments and equipment shall be disqualifying.

G. Maintaining Airways or ATC Routes.

(See Section II, IFR Navigation.)

H. Use of Radio Communications.

1. Description. The applicant may be asked to demonstrate the use of two-way radio voice communication procedures for reports, ATC clearances, and other instructions. Radio communications may be simulated at the discretion of the examiner.

2. Acceptable Performance Guidelines. The applicant's performance shall be evaluated on the basis of the use of the proper frequencies, correct phraseology, and the conciseness, clarity, and timeliness of transmissions. Acceptance of clearances based on facilities or frequencies not appropriate to the equipment being used or to the aircraft performance capabilities, shall be disqualifying.

I. Holding Procedures.

1. Description. The applicant may be directed, by ATC or the examiner, to hold in either a standard or a non-standard pattern at a specified fix. The applicant should make a proper entry as described in the *Airman's Information Manual*, remain within protected airspace, apply adequate wind correction, and accurately time the pattern so as to leave the fix at the time specified.

2. Acceptable Performance Guidelines. The applicant's performance shall be evaluated on compliance with instructions, entry procedure, orientation, accuracy and timing. Deviations of more than ±100 ft. from the prescribed altitude or more than ±10 kts. from holding airspeed shall be disqualifying.

J. Instrument Approach Procedures.

(See Pilot Operation III).

V. Emergencies

Objective: To determine that the applicant can promptly recognize and take appropriate action for abnormal or emergency conditions and equipment malfunctions while in instrument conditions.

A. Recovery from Unusual Attitudes.

1. Description. The examiner may place the airplane in unusual flight attitudes which may result from vertigo, wake turbulence, lapse of attention, or abnormal trim conditions. The applicant should recover and return to the original altitude and heading. For this demonstration, the examiner may limit the use of flight instruments by simulating malfunctions of the attitude indicator and heading indicator.

2. Acceptable Performance Guidelines. Evaluation shall be based on the promptness, smoothness, and accuracy demonstrated. All maneuvering shall be conducted within the operating limitations for the airplane used. Any loss of control which makes it necessary for the examiner to take over to avoid exceeding any operating limitation of the airplane shall be disqualifying.

B. Equipment or Instrument Malfunctions.

1. Description. The applicant may be asked to demonstrate the emergency operation of the retractable gear, flaps, and the electrical, fuel, deicing, and hydraulic systems if operationally feasible. Emergency operations such as the use of CO_2 pressure for gear extension, or the discharge of a pressure fire extinguisher system will be *simulated only*. Occasionally, during the performance of flight maneuvers described elsewhere in this guide, the examiner may simulate a partial or complete loss of flight instruments, navigation instruments, or equipment.

2. Acceptable Performance Guidelines. The applicant shall respond to emergency situations in accordance with procedures outlined in the manufacturer's published recommendations. The applicant's performance shall be evaluated on the basis of competency in maintaining aircraft control, knowledge of the emergency procedures, judgment displayed, and the accuracy of operations.

C. Loss of Radio Communications.

1. Description. The examiner may simulate loss of radio communications. The applicant should know the actions required pertaining to altitudes, routes, holding procedures, and approaches.

2. Acceptable Performance Guidelines. Evaluation shall be based on the applicant's knowledge of, and compliance with, the pertinent procedures required by Part 91 of the *Federal Aviation Regulations* and the emergency procedures outlined in the *Airman's Information Manual.* An explanation or simulation of the proper procedures for loss of radio communications is acceptable.

D. Engine-Out Procedures (Multiengine Airplane).

1. Description. The applicant may be asked to demonstrate the ability to positively and accurately maneuver the airplane after one engine has been throttled to simulate the drag of a feathered propeller, or with one propeller feathered, as agreed upon by the applicant and examiner. Feathering of a propeller for flight test purposes will be performed only under such conditions and at such altitudes and positions that a safe landing can readily be accomplished if an emergency develops or difficulty is encountered in unfeathering.

2. Acceptable Performance Guidelines. Evaluation shall be based on the applicant's ability to promptly identify the inoperative engine, and to follow the procedures outlined in the manufacturer's published recommendations. In cruising flight, the applicant shall maintain heading and altitude within ±20° and ±100 ft. If the airplane is incapable of maintaining altitude with an engine inoperative under existing circumstances, the applicant shall maintain an airspeed within ±5 kts. of the engine-out best rate-of-climb speed.

During approaches, the applicant shall promptly correct any deviation from the desired flight path.

Any loss of control that makes it necessary for the examiner to take over, or any attempt at prolonged flight contrary to the single—engine operating limitations of the airplane, shall be disqualifying.

E. Missed Approach Procedures.

1. Description. At any time during an instrument approach, the applicant may be asked to execute the missed approach procedure depicted on the approach chart being used. If the examiner does not specifically ask for the missed approach but fails to report the runway in sight at the DH on a precision approach, or the MAP (missed approach point) on a non-precision approach, the applicant should immediately initiate the missed approach procedure as described on the chart, or as directed by ATC.

2. Acceptable Performance Guidelines. Evaluation shall be made on the basis of the applicant's judgment in deciding when to execute the missed approach; the appropriateness of communications and navigation procedures; the ability to maintain positive airplane control, and to operate all airplane systems in accordance with applicable operating instructions for the airplane being used. Descent below the MDA or DH, as appropriate, prior to initiation of the missed approach procedure shall be disqualifying except in those instances where the runway environment was in sight at MDA or DH.

VI. Applicant's IFR Flight Test Checklist

The following is a typical list of items that are needed for an IFR flight test:

Properly Certificated Airplane with Dual Controls.

☐ View-Limiting Device.
☐ Aircraft Documents:
    Airworthiness Certificate.
    Registration Certificate.
    Operating Limitations.
☐ Weight and Balance Data
☐ Aircraft Maintenance Records:
    Airworthiness Inspections.
    Required Equipment Checks.
☐ FCC Station License.

Personal Equipment.

☐ En route Charts.
☐ SIDs and STARs.
☐ Instrument Approach Charts.

☐ Instrument Checklist.
☐ Current AIM.
☐ Flight Plan Form.
☐ Flight Logs.
☐ Computer.

Personal Records.

☐ Pilot Certificate.
☐ Medical Certificate.
☐ Signed Recommendation.
☐ Written Test Results.
☐ Logbook.
☐ Notice of Disapproval (if applicable).
☐ Approved School Graduation Certificate (if applicable).
☐ FCC Radiotelephone Operator Permit.
☐ Examiner's Fee (if applicable).

PART 4: THE COMMERCIAL PILOT FLIGHT TEST

(Extracted from the *Commercial Pilot-Airplane*, AC 61-55A.)

Use of the Flight Test Guide

The pilot operations listed in this flight test guide, indicated by Roman numerals, are required by Section 61.127 of Part 61 (revised) for issuance of a commercial pilot certificate. For the addition of a class rating to the pilot certificate, an applicant must pass a flight test appropriate to the pilot certificate held and applicable to the aircraft category and class rating sought.

This guide is intended only to outline appropriate pilot operations and the minimum standards for the performance of each procedure or maneuver which will be accepted by the examiner as evidence of the pilot's competency. It is not intended that the applicant be tested on every procedure or maneuver within each pilot operation, but only those considered necessary by the examiner to determine competency in each pilot operation.

When, in the judgment of the examiner, certain demonstrations are impractical (for example, malfunctions), competency may be determined by oral testing.

This guide contains an *Objective* for each required pilot operation. Under each pilot operation, pertinent procedures or maneuvers are listed with *Descriptions* and *Acceptable Performance Guidelines*.

1. The *Objective* states briefly the purpose of each pilot operation required on the flight test.

2. The *Description* provides information on what may be asked of the applicant regarding the selecting procedure or maneuver. The procedures or maneuvers listed have been found most effective in demonstrating the objective of that particular pilot operation.

3. The *Acceptable Performance Guidelines* include the factors which will be taken into account by the examiner in deciding whether the applicant has met the objective of the pilot operation. The airspeed, altitude, and heading tolerances given represent the minimum performance expected in good flying conditions. However, consistently exceeding these tolerances before corrective action is initiated is indicative of an unsatisfactory performance. Any procedure or action, or the lack thereof, which requires the intervention of the examiner to maintain safe flight will be disqualifying. Failure to exercise proper vigilance or to take positive action to ensure that the flight area has been adequately cleared for conflicting traffic will also be disqualifying.

Emphasis will be placed on procedures, knowledge, and maneuvers which are most critical to a safe performance as a pilot. The demonstration of prompt stall recognition, adequate control, and recovery techniques will receive special attention. Other areas of importance include spatial disorientation, collision avoidance, and wake turbulence hazards.

The applicant will be expected to know the meaning and significance of the airplane performance speeds important to the pilot, and be able to readily determine those speeds for the airplane used for the flight test. These speeds include:

V_{so} —the stalling speed or minimum steady flight speed in landing configuration.

V_y —the speed for the best rate of climb.

V_x —the speed for the best angle of climb.

V_a —the design maneuvering speed.

V_{ne} —the never exceed speed.

$V_{le.}$ —the maximum landing gear extended speed.

$V_{fe.}$ —the maximum flap extended speed.

If the commercial pilot flight test is taken in a multiengine airplane, the following additional performance speeds are also applicable:

V_{mc} —the minimum control speed with the critical engine inoperative.

The speed for the best angle-of-climb with one engine inoperative.

The speed for the best rate-of-climb with one engine inoperative.

A multiengine class rating issued on the basis of a flight test in a multiengine airplane which has no engine-out minimum control speed will bear the notation LIMITED TO CENTER THRUST. Such airplanes typically have jet engines in or on the fuselage, or reciprocating engines in tandem on the centerline of the fuselage.

In the event the applicant takes the instrument pilot flight test and the commercial pilot flight test simultaneously, the maneuvers selected from this guide and AC 61-56, *Instrument Pilot Flight Test Guide*, may be combined and evaluated together, where practicable.

General Procedures for Flight Tests

The ability of an applicant for a commercial pilot certificate, or for an aircraft class, or instrument rating on that certificate, to perform the required pilot operations is based on the following:

1. Executing procedures and maneuvers within the aircraft's performance capabilities and limitations, including use of the aircraft's systems.

2. Executing emergency procedures and maneuvers appropriate to the aircraft.

3. Piloting the aircraft with smoothness and accuracy.

4. Exercising judgment.

5. Applying aeronautical knowledge.

6. Showing masterful handling of the aircraft, with the successful outcome of a procedure or maneuver never seriously in doubt.

If the applicant fails any of the required pilot operations, the flight test is failed. The examiner or the applicant may discontinue the test at any time when the failure of a required pilot operation makes the applicant ineligible for the certificate or rating sought. If the test is discontinued the applicant is entitled to credit for only those entire pilot operations that were successfully performed.

Flight Test Prerequisites

An applicant for the commerical pilot flight test is required by the revised Part 61 to have: (1) a private pilot certificate with an airplane rating or meet the flight experience required for a private pilot certificate (airplane rating) and pass the applicable (Part 61, Subpart D) written and practical test; (2) an instrument rating (airplane) or the following limitation will be placed on the commercial pilot certificate "carrying passengers in airplanes for hire is prohibited at night and on cross-country flights of more than 50 nautical miles"; (3) passed the appropriate commercial pilot written test within 24 months before the date the flight test is taken; (4) the applicable instruction and aeronautical experience prescribed (Part 61 Subpart E) for a commercial pilot certificate; (5) a first or second class medical certificate issued within the past 12 months; (6) reached at least 18 years of age; and (7) a written statement from an appropriately certificated flight instructor certifying that the applicant has been given flight instruction in preparation for the flight test within 60 days preceding the date of application, and was found competent to pass the test and to have a satisfactory knowledge of the subject areas in which a deficiency was shown on the airman written test report.

Airplane and Equipment Requirements for Flight Test

The applicant is required by revised Section 61.45 to provide an airworthy airplane for the flight test. This airplane must be capable of, and its operating limitations must not prohibit, the pilot operations required on the test. The following equipment is relevant to the pilot operations required by revised Section 61.127 for the commercial pilot flight test:

1. Two way radio suitable for voice communications with aeronautical ground stations.

2. A radio receiver which can be utilized for available radio navigation facilities (may be the same radio used for communications).

3. If the applicant does not hold an instrument rating and takes the instrument pilot flight test simultaneously with the commercial pilot flight test, the airplane used must have appropriate flight instruments for the control of the airplane during instrument conditions. Appropriate flight instruments are considered to be those outlined in FAR Part 91 for flight under instrument flight rules. However, if the instrument privileges associated with the commercial pilot certificate are not desired, the applicant is not required to provide for the flight test an airplane equipped for flight under instrument flight rules.

4. Engine and flight controls that are easily reached and oper-

ated in a normal manner by both pilots, unless the examiner determines the flight test can be conducted safely without them. Fully functioning dual controls are required by Part 91 for simulated instrument flight.

5. Operating instructions and limitations. The applicant should have an appropriate checklist, an Owner's Manual/Handbook, or if required for the airplane used, an FAA approved Airplane Flight Manual. Any operating limitations or published recommendations of the manufacturer that are applicable to the specific airplane shall be observed.

6. An airplane equipped with retractable landing gear, flaps, and controllable propeller(s), if a landplane rating is sought.

I. Preflight Duties

Objective: To determine that the applicant can ensure meeting the requirements to act as pilot in command, that the airplane is airworthy and ready for safe flight, and that suitable weather conditions exist for the proposed flight.

A. Certificates and Documents

1. Description. The applicant may be asked to present his pilot and medical certificates and to produce and explain the airplane's registration certificate, airworthiness certificate, operating limitations or FAA-approved Airplane Flight Manual (if required), airplane equipment list, and required weight and balance data. In addition, the applicant is expected to explain the airplane and engine logbooks or other maintenance records.

2. Acceptable Performance Guidelines. The applicant shall be knowledgeable regarding the location, purpose, and significance of each required item.

B. Airplane Performance and Limitations.

1. Description. The applicant may be orally quizzed on the performance capabilities and approved operating procedures and limitations of the airplane used. This includes power settings, placarded speeds, and fuel and oil requirements. In addition, the manufacturer's published recommendations[2] or FAA-approved

---

[2] The phrase "manufacturer's published recommendations" is used hereafter in this guide to denote FAA approved Airplane Flight Manual material when such material has been approved for the airplane type or other manufacturer's published recommendations, such as "Owner's Manual," "Owner's Handbook," "Bulletins," and "Letters" for the safe operation of the airplane model or series, in the absence of an approved Airplane Flight Manual.

Airplane Flight Manual should be used to determine the effects of temperature, pressure altitude, wind, and gross weight of the airplane's performance.

2. Acceptable Performance Guidelines. The applicant shall obtain, explain, and apply the information which is essential in determining the performance capabilities and limitations of the airplane used.

C. Weight and Balance.

1. Description. The applicant may be asked to demonstrate the application of the approved weight and balance data for the airplane used to determine that the gross weight and center of gravity (C.G.) location are within allowable limits. Charts and graphs provided by the manufacturer may be used.

2. Acceptable Performance Guidelines. The applicant shall determine the empty weight, C.G., maximum allowable gross weight, useful load (fuel, passengers, baggage) by reference to appropriate publications, and shall apply that information to determine that the actual gross weight and C.G. are within approved limitations.

D. Weather Information.

1. Description. The applicant may be asked to obtain Aviation Weather Reports, Area and Terminal Forecasts, and Winds Aloft Forecasts pertinent to the proposed flight.

2. Acceptable Performance Guidelines. The applicant shall demonstrate ability to select or determine weather information that is pertinent and how to best obtain that information, and to interpret and understand its significance with respect to the proposed flight.

E. Line Inspection.

1. Description. The applicant may be asked to demonstrate a visual inspection to determine the airplane's airworthiness and readiness for flight. This includes all required equipment and documents. A checklist provided by the manufacturer or operator should be used.

2. Acceptable Performance Guidelines. The applicant shall use an orderly procedure in conducting a preflight inspection of the airplane and shall know the significance of each item checked and recognize any unsafe condition.

F. Airplane Servicing.

1. Description. The applicant may be asked to demonstrate a visual inspection to determine that the fuel is of the proper grade and type and that the fuel and oil supply is adequate for the proposed flight. Appropriate action should be taken to eliminate possible fuel contamination in the airplane. In addition, the applicant may be asked to check the adequacy of the oxygen supply on airplanes so equipped.

2. Acceptable Performance Guidelines. The applicant shall know the grade and type of oil and fuel specified for the airplane and be able to determine the amount of oxygen and fuel required to complete the flight. Applicant shall know the proper steps for avoiding fuel contamination during and following servicing.

G. Engine and Systems Preflight Check.

1. Description. The applicant may be asked to demonstrate a check to determine that the engine is operating within acceptable limits and that all systems, equipment, and controls are functioning properly and adjusted for takeoff. A checklist provided by the manufacturer or operator should be used.

2. Acceptable Performance Guidelines. The applicant shall use proper procedures in engine starting and runup and in checking airplane systems, equipment, and controls to determine that the airplane is ready for flight. Careless operation in close proximity to obstructions, ground personnel, or other aircraft shall be disqualifying.

II. Flight at Critically Slow Airspeeds

Objective: To determine that the applicant can competently maneuver the airplane at critically slow airspeeds in various attitudes and configurations, and can recognize imminent stalls and accomplish prompt, effective recoveries in all normally anticipated flight situations.

A. Maneuvering at Minimum Controllable Airspeeds.

1. Description. The applicant may be asked to maneuver at such an airspeed that controllability is minimized to the point that if the angle of attack or load factor is further increased, an immediate stall would result. The maneuver should be accomplished in medium-banked level, climbing and descending turns, and straight-and-level flight with various flap settings in both cruising and landing configurations.

2. Acceptable Performance Guidelines. The applicant shall be evaluated on competence in establishing the minimum controllable airspeed, in positively controlling the airplane, and in recognizing incipient stalls. Primary emphasis shall be placed on airspeed control. During straight-and-level flight at this speed, the applicant shall maintain altitude within ±50 ft. and heading within ±10° of that assigned by the examiner. Inadequate surveillance of the area prior to and during the maneuver, or an unintentional stall shall be disqualifying.

B. Imminent Stalls.

1. Description. The applicant may be asked to demonstrate recoveries from imminent stalls entered from straight flight and from turning flight with power on and with power off. Applicant is expected to place the airplane in the attitude and configuration appropriate for flight situations such as takeoffs, departures, landing approaches, and accelerated maneuvers, as directed by the examiner. The applicant should apply control pressures which result in an increase in angle of attack until the first buffeting or decay of control effectiveness is noted. The recovery should be accomplished immediately by reducing the angle of attack. Recoveries should be made with or without power, as directed by the examiner.

2. Acceptable Performance Guidelines. The applicant shall be evaluated on competence in recognizing the indications of an imminent stall and in taking prompt, positive control action to prevent a full stall. The applicant shall be disqualified if a full stall occurs or if it becomes necessary for the examiner to take control of the airplane to avoid excessive airspeed, excessive loss of altitude, or a spin.

III. Takeoffs and Landings

Objective: To determine that the applicant is competent in performing takeoffs and landings under all normally anticipated conditions in a landplane equipped with retractable landing gear, flaps, and controllable propeller(s), or in a seaplane with flaps and controllable propeller.

A. Normal and Crosswind Takeoffs (Landplanes).

1. Description. The applicant may be asked to demonstrate normal and crosswind takeoffs by aligning the airplane with the runway or takeoff surface and applying takeoff power smoothly and positively while maintaining directional control. In crosswind takeoffs, the applicant is expected to hold aileron into the wind and maintain a straight takeoff path by use of rudder and to gradually establish a pitch attitude which produces an angle of attack that permits normal acceleration and lift-off.

The applicant may be asked to make at least one crosswind takeoff with sufficient crosswind to require the use of crosswind techniques, but not in excess of the crosswind limitations of the airplane used.

2. Acceptable Performance Guidelines. The applicant's competence in performing normal and crosswind takeoffs shall be evaluated on power application, smoothness, wind drift corrections, coordination, and directional control. The applicant shall maintain climb speed within ±5 knots of the desired initial climb speed after lift-off. Improper or incomplete after-takeoff procedures shall be disqualifying.

B. Normal and Crosswind Landings (Landplanes).

1. Description. The applicant may be asked to accomplish normal and crosswind landings using a final approach speed equal to 1.3 times the power-off stalling speed in landing configuration (1.3 V_{so}), or the final approach speed prescribed by the manufacturer. The landings may be accomplished with or without power, with touchdowns being made within the area specified by the examiner. Landings may be made with full flaps, partial flaps, or no flaps. Forward slips and a slip to a landing may be performed with or without flaps, unless prohibited by the airplane's operating limitations.

In a tailwheel type airplane, the main wheels and tailwheel should touch the runway simultaneously at or near power-off stalling speed. In a nosewheel type airplane, the touchdown should be on the main wheels with little or no weight on the nosewheel. In strong gusty surface winds, in a tailwheel type airplane, the round-out should be made to an attitude which permits touchdown on the main wheels only. In crosswind conditions, wind drift corrections should be made throughout the final approach and touchdown. Adequate corrections and positive directional control should be maintained during the after-landing roll.

The applicant may be asked to make at least one crosswind landing with sufficient crosswind to require the use of crosswind techniques but not to exceed the crosswind limitations of the airplane. The applicant may be asked to demonstrate at least one power-off accuracy landing.

2. Acceptable Performance Guidelines. The applicant's competence in performing normal and crosswind landings shall be evaluated on the basis of landing technique, judgment, wind drift correction, coordination, power technique, and smoothness. Proper final approach speed shall be maintained within ±5 knots, and touchdown accomplished in the proper landing attitude beyond and within 200 ft. of a line or mark specified by the examiner.

Improper or incomplete pre-landing procedures, touching down with an excessive side load on the landing gear, and poor directional control shall be disqualifying.

F. Wake Turbulence Avoidance.

1. Description. The applicant may be asked to explain how, where, and when wingtip vortices are generated, their characteristics and associated hazards, and recommended courses of action to remain clear of those hazards.

2. Acceptable Performance Guidelines. The applicant shall identify the conditions and locations in which wingtip vortices may be encountered and adjust the flightpath in a manner to avoid those areas. Failure to follow recommended procedures for minimizing the likelihood of flying into wingtip vortices shall be disqualifying.

IV. Maximum Performance Maneuvers

Objective: To determine the applicant's competence in obtaining maximum airplane performance.

A. Short Field Takeoff and Maximum Climb.

1. Description. The applicant may be asked to demonstrate a takeoff which is used at short fields with obstacles. Applicant is expected to use prompt, smooth application of power and rotate to lift-off just as the best angle-of-climb airspeed is attained. Applicant should maintain that speed until the assumed obstacles have been cleared. The applicant is expected to know and understand the effectiveness of the best angle-of-climb and best rate-of-climb airspeeds of the airplane to obtain maximum climb performance. The flap settings and airspeed prescribed by the airplane manfacturer should be used.

2. Acceptable Performance Guidelines. The applicant's competence shall be evaluated on planning, smoothness, directional control, and accuracy. In simulating a short field takeoff, the lift-off and climb shall be performed within ±5 knots of the best angle-of-climb speed. The applicant shall obtain performance which is comparable to that presented in the airplane's performance data. Improper flap or propeller setting or premature retraction of the landing gear shall be disqualifying.

B. Short Field Approach and Landing.

1. Description. The applicant may be asked to demonstrate a landing from over an assumed 50-ft. obstacle. Final approach speed should result in little or no floating after the throttle is closed during the flare for touchdown. Touchdown should be made within the area designated by the examiner, at minimum controllable airspeed, in approximately the pitch attitude which results in a power-off stall. Upon touchdown in landplanes, the applicant is expected to properly apply brakes and/or reverse thrust to minimize the afterlanding roll. Power, flaps, or moderate slips should be used as necessary on the last segment of the final approach.

2. Acceptable Performance Guidelines. Performance shall be evaluated on the basis of planning, coordination, smoothness, and accuracy. The applicant shall accurately control the angle of descent and airspeed on final approach so that floating is minimized during the flare. After touchdown, the airplane should be brought smoothly to a stop within the shortest possible distance consistent with safety. Improper or incomplete prelanding procedures, touching down with an excessive side load on the landing gear, or poor directional control shall be disqualifying.

C. Soft Field Takeoff.

1. Description. The applicant may be asked to demonstrate a soft field takeoff. This should be accomplished with the wing at a relatively high angle of attack so as to transfer the weight from the wheels to the wing as soon as possible. As soon as the elevators become effective, a positive angle of attack should be established to lighten the load on the nosewheel or the tailwheel. After becoming airborne, the pitch attitude should be adjusted with the wheels just clear of the surface to allow the airplane to accelerate. Care should be exercised to prevent settling back to the ground. As the airplane reaches best angle-of-climb or best rate-of-climb speed, whichever is appropriate for the field conditions, adjust the pitch attitude to maintain the desired climb speed. The flap setting used should be in accordance with the manufacturer's recommendations.

2. Acceptable Performance Guidelines. The applicant's performance shall be evaluated on planning, directional control, smoothness, and accuracy. The applicant shall lift off at a speed that does not exceed the power-off stalling speed and maintain the proper climb speed within ±5 knots. Improper flap or propeller setting or premature retraction of the landing gear shall be disqualifying.

D. Soft Field Landing.

1. Description. The applicant may be asked to demonstrate a simulated soft field landing from a normal approach with touchdown at the slowest possible airspeed to permit the softest possible touchdown and a short landing roll. A nose-high attitude should be maintained during the roll-out and the flaps promptly retracted (if recommended by the manufacturer) to prevent damage from mud or slush thrown by the wheels.

2. *Acceptable Performance Guidelines.* The applicant's performance shall be evaluated on planning, smoothness, and accuracy. Final approach airspeed within ±5 knots of that prescribed shall be maintained. *During flap retraction the applicant shall exercise extreme caution and maintain positive control.* Improper or incomplete prelanding procedures, touching down with an excessive side load on the landing gear, or poor directional control shall be disqualifying.

E. Chandelles.

1. *Description.* The applicant may be asked to perform chandelles both to the left and to the right. The maneuver should be entered at maneuvering speed by establishing an appropriate angle of bank and pitch attitude which will produce a maximum climbing turn. A coordinated recovery should be started at 90° of turn and continued so that the 180° point is reached with the wings level and the airspeed just above a stall. The pitch attitude should then be lowered to the level flight attitude for the existing airspeed. The maneuver may be accomplished with the use of a fixed power setting or the coordinated use of power.

2. *Acceptable Performance Guidelines.* Evaluation of performance shall be based on planning, airspeed control, coordination, smoothness, and orientation. The applicant shall complete the chandelles within ±10° of the desired heading, and recover with an airspeed not more than 5 knots above stalling speed. A stall during the maneuver shall be disqualifying.

F. Lazy Eights.

1. *Description.* The applicant may be asked to perform a lazy eight. This consists of two 180° turns in opposite directions, with a symmetrical climb and dive performed during each turn. The airplane should be constantly rolled from one bank to the other, while the pitch attitude is constantly changed from climbs to dives. The loops should be symmetrical with portions above and below the horizon equal in size. At no time during the maneuver should the airplane attitude, control positions, or control forces be held constant.

2. *Acceptable Performance Guidelines.* The applicant's performance shall be evaluated on planning, coordination, smoothness, attitude, and airspeed control. A persistent gain or loss of altitude at the completion of each lazy eight, or repeated slipping or skidding, shall be disqualifying.

G. Steep Power Turns.

1. *Description.* The applicant may be asked to execute maximum performance turns in either direction using a bank of at

least 50°, gradually imposing relatively high load factors well within the structural limits of the airplane. Power should be increased as the bank is established and decreased during the roll-out as required to maintain altitude and airspeed.

2. Acceptable Performance Guidelines. The applicant's competence shall be evaluated on planning, coordination, smoothness, prompt stabilization of the turns, and orientation during the maneuver. Variations of more than ±100 ft. from the entering altitude should be disqualifying.

H. Descents.

1. Description. The applicant may be asked to establish the airplane configuration and glide speed which will result in the greatest forward distance with the least loss of altitude.

2. Acceptable Performance Guidelines. The applicant shall promptly establish and maintain the proper configuration and pitch attitude which will produce the most efficient speed for glides.

I. Steep Spirals.

1. Description. The applicant may be asked to perform a steep spiral around a selected ground reference point and continue for a minimum number of turns specified by the examiner. Recovery should be made at a specified point relative to the ground reference. A constant radius around the point should be maintained by varying the bank to correct for wind effect.

2. Acceptable Performance Guidelines. The applicant shall be competent in entering, maintaining, and recovering from steep spirals using smooth coordinated controls. Loss of orientation, descending below a safe altitude, or excessive variation of pitch attitude shall be disqualifying. Observance of the following limits will be accepted as competent performance:

    a. Airspeed within ±10 knots of that recommended.

    b. Steepest bank between 50° and 55°.

    c. Recovery at the specified point or at a safe altitude.

    d. Uniform radius around the reference point.

V. Operation of Airplane Equipment

Objective: To determine that the applicant has a thorough knowledge of, and can competently perform normal and emergency operations of all systems and equipment for the airplane used on the flight test.

A. Retractable Landing Gear
Operation (Landplane or Amphibian).

1. Description. The applicant may be asked to demonstrate the normal and emergency operation of the landing gear in accordance with the manufacturer's published recommendations.

2. Acceptable Performance Guidelines. Performance shall be evaluated on the applicant's knowledge of normal and emergency procedures and the accuracy of operations. Untimely operation of the landing gear which creates a hazard shall be disqualifying.

B. Flap Operation.

1. Description. The applicant may be asked to demonstrate the normal and emergency operation of flaps in accordance with the manufacturer's published recommendations. Applicant should be knowledgeable regarding the effect of flaps on airplane performance.

2. Acceptable Performance Guidelines. Performance shall be evaluated on the applicant's knowledge of normal and emergency procedures and the accuracy of the operations. Untimely operation of the flaps which creates a hazard shall be disqualifying.

C. Controllable Propeller.

1. Description. The applicant may be asked to demonstrate the operation of the propeller in accordance with the manufacturer's published recommendations. Applicant should be knowledgeable regarding the effect of throttle and propeller settings on airplane performance.

2. Acceptable Performance Guidelines. Performance shall be evaluated on the applicant's knowledge of operating procedures and the accuracy of operation. Improper propeller settings which may damage the engine or create a hazard shall be disqualifying.

VI. Emergency Procedures

Objective: To determine that the applicant has a thorough knowledge of, and can competently perform emergency procedures for all systems and equipment installed in the airplane used on the flight test.

A. Power Loss.

1. Description. The applicant may be asked to demonstrate knowledge of corrective actions for: (1) partial loss of power, (2) complete power failure, (3) rough engine, (4) carburetor/induction system ice, and (5) fuel starvation. The examiner will, with no advance warning, reduce the power to simulate engine malfunction.

2. Acceptable Performance Guidelines. The applicant shall be able to immediately recognize the loss of power and take prompt remedial action, and shall use good judgment and techniques to minimize the danger to occupants and the airplane. The applicant shall perform emergency procedures for loss of power in compliance with the manufacturer's published recommendations. Any action which creates an unnecessary hazard shall be disqualifying.

B. Equipment Malfunctions.

1. Description. The applicant may be asked to demonstrate knowledge of corrective actions for (1) inoperative electrical system, (2) inoperative hydraulic system, (3) gear and flap malfunctions, (4) door opening in flight, and (5) inoperative elevator trim tab. Applicant may be asked, when practicable, to actually perform the proper remedial action for such emergency conditions as flap malfunctions, landing gear malfunctions, or inoperative electrical or hydraulic system.

2. Acceptable Performance Guidelines. Performance shall be evaluated on the applicant's prompt analysis of the situation and performance of emergency procedures in conformance with the manufacturer's published recommendations. Any action which creates unnecessary additional hazards shall be disqualifying.

C. Fire in Flight.

1. Description. The applicant is expected to recognize the symptoms of electrical fires and fuel fires. When the examiner describes the symptoms of a fire situation, the applicant is expected to follow emergency procedures appropriate for combating the type of fire.

2. Acceptable Performance Guidelines. The applicant shall be able to recognize the type of fire described, determine its location, and explain the proper procedure for extinguishing the fire or for safely terminating the flight.

D. Collision Avoidance Precautions.

1. Description. The applicant is expected to exercise conscientious and continuous surveillance of the airspace in which the airplane is being operated to guard against potential mid-air collisions. In addition to "see and avoid" practices, applicant is expected to use VFR Advisory Service at nonradar facilities, Airport Advisory Service at nontower airports or FSS locations, and Radar Traffic Information Services where available.

2. Acceptable Performance Guidelines. The applicant shall maintain continuous vigilance for other aircraft and take immediate actions necessary to avoid any situation which could result in a mid-air collision. Extra precautions shall be taken, particularly in areas of congested traffic, to ensure that other aircraft are are not obscured by the aircraft's structure. When traffic advisory service is used, the applicant shall understand terminology used by the radar controller in reporting positions of other aircraft. Failure to maintain proper surveillance shall be disqualifying.

VII. Applicants' Commercial Pilot Flight Test Checklist

The following is a typical list of items that are required for the Commercial Pilot flight test:

Acceptable Airplane.

☐ Aircraft Documents:
   Airworthiness Certificate.
   Registration Certificate.
   Operating Limitations.
   Weight and Balance Data.

☐ Aircraft Maintenance Records:
   Airworthiness Inspections.

☐ FCC Station License.

Personal Records.

☐ Pilot Certificate.

☐ Medical Certificate.

☐ Signed Recommendation (if applicable).

☐ Written Test Report (AC Form 8060-37).

☐ Logbook with Instructors Endorsement.

☐ Notice of Disapproval (if applicable, FAA Form 8060-5).

☐ Approved School Graduation Certificate (if applicable).

☐ FCC Radiotelephone Operator Permit.

☐ Examiner's Fee (if applicable).

Chapter 10
The Biennial Flight Review

The Biennial Flight Review is required of *all* pilots. As set forth in FAR 61.57, the purpose of the review is to assume that the pilot can *safely* exercise the privileges of his certificate. The review can be given by a flight instructor, a company check pilot, or a military check pilot as well as by the FAA. The passage of a pilot proficiency check in pursuit of a new rating also constitutes a Biennial Flight Review.

Unlike the pilot medical, the Biennial is *not* valid to the end of the calendar month in which it was taken; its valid time extends to the *same date* two years hence from the date it was taken.

To define the content of a Biennial Flight Review and establish criteria for such, a committee consisting of the following aviation organizations was formed:

AOPA	NAP
EAA	NAFI
GAMA	Ohio State Univ., Dept. of Aviation
NATA	FAA, General Aviation Division

This committee developed an excellent set of guidelines for the conduct of the biennial flight review. The remainder of this chapter contains the recommendations of the committee.

THE BFR CONCEPT

The Biennial Flight Review is a cooperative endeavor to provide the pilot with a periodic assessment of his flying skills and to determine if there has been deterioration in areas which may reasonably affect his safety. The BFR should be a currency evaluation accomplished in an economical and expeditious manner, and at the same time, provide a learning situation rather than a "check flight" atmosphere.

The character of any Biennial Flight Review should be established in a prereview discussion between the pilot and flight instructor. The BFR's basic character, including the elements to be covered in both the oral and flight portions, should be understood by both the pilot and flight instructor prior to initiating any phase of the review. The principal point of these guidelines is to reinforce the attitude that each Biennial Flight Review will be unique to each pilot/instructor combination, and that uniqueness will be the product of the prereview dialogue between the pilot and the instructor.

The Biennial Flight Review is not a "test" or a "check ride", it is a review where assistance and instruction may be given as necessary to improve the pilot's demonstrated performance and assist in the satisfactory completion of the review. The availability and extent of dual instruction provided during the flight should be determined during the prereview discussion.

Each Biennial Flight Review should be individually tailored to meet, in the reasonable discretion of the reviewing flight instructor, the basic safe operating demands of the pilot being reviewed. The primary objective of any BFR should be to assess the pilot's ability to successfully perform, and be knowledgeable of, safe flight operations. Rather than using standard guidelines or a list of maneuvers, flight instructors are encouraged to determine the safe operating needs of each pilot, and then formulate a meaningful Biennial Flight Review tailored to meet those needs. The review should assess the pilot's broad awareness of applicable regulations, procedures, and good operating practices as opposed to committing rote specifies to memory. Finally, reviewing flight instructors should not make the oral and flight reviews overly demanding in terms of the total number of operations, nor should they require perfection in those subject areas and operations evaluated during those reviews.

THE PILOT PROFILE

The first step in the conduct of a Biennial Flight Review should incorporate a review of the pilot's background. This initial discussion portion of the BFR will serve two basic purposes: First, it provides both pilot and instructor with an opportunity to assess each other. Additionally, it gives both parties a chance to discuss individual experience, recent flight experience, and what each person expects to encounter and gain during the BFR. Second, the "pilot profile" session serves the purpose of providing a point in the review process to make sure that all necessary paperwork is in order; this should also include all documents necessary for the operation of the aircraft.

During this phase of the BFR the flight instructor should review the pilot's flight experience to provide a basis for being of most assistance to that pilot. Following the assessment of flight experience the instructor can begin to formulate the character of the oral and flight review most appropriate for that pilot. At this point it should be possible for the reviewing flight instructor to provide the pilot with an estimate of the approximate length of time that will be required to accomplish the flight portion of the BFR. Since no two Biennial Flight Reviews will be the same, each pilot and flight instructor for each particular BFR should recognize that this review will be designed specifically for the person being evaluated based on the pilot's individual characteristics and qualifications, as well as the nature of the aircraft involved in the Biennial Flight Review.

This planning phase of a Biennial Flight Review may be accomplished prior to the actual oral and flight review. Preliminary discussion can provide an opportunity for the pilot to correct any paperwork problems; furthermore, it gives the flight instructor a chance to suggest regulations to review, books to read, and manuals and charts to obtain, if necessary. While Biennial Flight Reviews are adaptable to segmented sessions, users of this approach should continue to bear in mind that BFR's are also intended to be accomplished in an economical and expeditious manner.

Generally speaking, this stage of the BFR can be of practical value to both individuals for at least one other significant reason. During this phase both the pilot and the instructor can begin to determine, and probably even decide, if either of them wants to proceed any further with this particular review. If it appears that there may be discernible, or potentially irreconcilable, conflicts of philosophies, personalities, etc.; this is the time to consider a different instructor, fixed base operator, pilot, or just another time.

REVIEW OF APPLICABLE RULES

The second step in any pilot's Biennial Flight Review should involve a review of applicable Federal Aviation Regulations. This review must encompass those operational and flight provisions of FAR Part 91 appropriate to the operations of the pilot receiving the

BFR. As with the actual flight review, this portion of the Biennial Flight Review should not be conducted on a strictly test basis, rather it is an opportunity for the reviewing instructor to assist the pilot through a discussion of regulations and their relationship to operational safety. Prior to or following this phase of the review, flight instructors may want to refer pilots to the appropriate "Advisory Circular Checklist and Status of Federal Aviation Regulations" notice and various other materials as reference sources for additional study and review.

PREFLIGHT PROCEDURES REVIEW

The preflight procedures segment of a Biennial Flight Review should include an assessment of all those activities which the pilot would normally be expected to engage in prior to actually starting the engine of the aircraft. This phase of the Review could include, but not necessarily be limited to, an analysis of current and forecast weather, flight planning procedures, and preflight of the aircraft in general and for the particular flight anticipated. A review of fuel considerations, weight and balance computations, as well as performance and navigation charts appropriate for the flight could all be included at this point in the evaluation.

While it is certainly not mandatory, the planning of a cross-country flight could be included to illustrate the practical advantages of obtaining weather information, planning a suitable route and altitude, estimating elapsed time, determining fuel required and allowable, and completing weight and balance computations.

In this preflight procedures phase, as in all other phases of the BFR, the instructor should render assistance by questioning, correcting and instructing rather than testing.

BASIC FLIGHT REVIEW

The purpose of the flight portion of the Biennial Flight Review is to permit the flight instructor to observe and evaluate those flight operations necessary for a review of a pilot's habits, skills and procedures. It is not intended to be a critique of the pilot's ability to execute specific maneuvers such as those found in flight training or in testing for certificate or rating qualifications.

The objectives of the flight segment of a Biennial Flight Review can be accomplished through the pilot's demonstration of, but not limited by, such operational activities as preflight procedures, airport and traffic pattern operations, abnormal (cross-wind and short-field landings and take-offs, etc.) and emergency procedures (including inadvertent weather penetration).

The actual selection of flight operations and procedures the pilot is asked to demonstrate should be left to the discretion of the reviewing flight instructor; thereby enabling him to tailor the flight review to the needs of the individual pilot. Emphasis on overall safety of flight operations should be stressed more than precision execution of pure training maneuvers. Furthermore, since there will seldom be time in a Biennial Flight Review to evaluate all flight operations, it is recommended that particular attention be given to those operations that seem to cause greatest difficulty to the pilot being reviewed or have the greatest tendency to get most pilots into trouble.

Both the pilot and reviewing flight instructor should be aware that Biennial Flight Reviews need not be limited to evaluation only, but may also be instructional. That intent can be realized if, as a product of the initial pilot/instructor discussion session, the instructor agrees to provide instruction assistance as necessary when both he and the pilot identify "weak areas" in need of additional practice. It is generally accepted that a reasonable number of instructor performed "demonstrations" of proper technique are a realistic expectation following an initial marginal or unsatisfactory pilot-performed flight operation.

Holders of advanced ratings and certificates may desire an evaluation of their capabilities and skills not included in the basic oral and flight reviews; such evaluations are optional, but pilots are encouraged to seek maximum benefits from the Biennial Flight Review process. During the pre-review or "pilot profile" phases of the BFR, flight instructors may want to point out the advisability of a more comprehensive review if such advanced ratings and skills are currently relied on during flights typical of that pilot's normal operations.

Evaluation of the total flight operations should be made to determine a pilot's satisfactory performance, and that evaluation should be made on the basis of a simple satisfactory vs. unsatisfactory system. The word "satisfactory" is used even though a Biennial Flight Review is not an official flight check. The term is used only to provide the reviewer with a minimum standard baseline on which to base a decision and comments regarding the review.

POSTFLIGHT DISCUSSION AND RECOMMENDATION

This phase of the Biennial Flight Review is potentially the

most important part of the entire evaluation process because it is here that the reviewing flight instructor gets a chance to discuss with the pilot what the entire review has revealed. In order for the BFR to be of any real value, the pilot must realize the importance of listening to this appraisal of his skills with an open mind, not with one closed to any hint of criticism. On the other hand, the person giving the appraisal must make a real effort to provide a clear and constructive debriefing of the BFR which has just taken place.

All Biennial Flight Reviews, whether accomplished satisfactorily or not, should be concluded with a helpful, positive discussion and suggestions for any remedial or improvement actions that the instructor considers beneficial. This postflight discussion should provide an honest, objective, and lucid appraisal of the pilot's current ability to successfully perform, and be knowledgeable of, safe flight operations, at least to the extent that such abilities were capable of evaluation during this particular Biennial Flight Review.

The logbook entry is the proper form of proof of compliance with the Biennial Flight Review requirement. There should be no endorsement, or any indication of any kind, in the pilot's logbook reflecting the unsatisfactory nature of a Biennial Flight Review; nor should a satisfactory endorsement allude to any unsatisfactory part of the BFR. If, in the opinion of the reviewing instructor, the pilot's Biennial Flight Review performance cannot be considered as having been satisfactory, no logbook entry will be made; furthermore, the instructor should recommend the appropriate remedial action necessary to bring the pilot up to an acceptable level of performance or knowledge. At this point, if the BFR was unsatisfactory, the pilot has the option of continuing with that instructor or choosing another instructor for review, assistance, or an attempt as another complete Biennial Flight Review.

Pilot/Controller Glossary and Basic IFR Clearance Shorthand

This glossary is an abbreviated list of commonly used VFR and IFR aviation terms. A degree of emphasis has been placed on IFR terms in view of the continuing increase in the complexity of IFR flight and the significance of knowing exactly what each IFR term implies. The selection of words and definitions is from the *Airman's Information Manual* (AIM), *Basic Flight Information and ATC Procedures Volume,* and also FAR Part I, *Definitions and Abbreviations*. In certain instances, published definitions have been shortened for ease of reading while in other cases, material has been added to increase clarity. For additional information, you may wish to subscribe to the AIM. Volume 1, *Basic Flight Information and ATC Procedures*, contains information on Safety of Flight, Medical Facts for Pilots, ATC Procedures, Emergency Procedures, and many related items. This volume of the AIM may be obtained from the Superintendent of Documents, Government Printing Office, Washington, D.C. 20402. The current price for a one-year subscription is $5.00 (two issues). FAR Part 1 is also available from the Superintendent of Documents as well as at most aviation book stores.

COMMON VFR AND IFR AVIATION TERMS

abeam—An aircraft is *abeam* a fix, point or object when that fix, point or object is approximately 90 degrees to the right or left of the aircraft track. Abeam indicates a general position rather than a precise point.

abort—To terminate a preplanned aircraft maneuver.

acknowledge—"Let me know that you have received and understand my message."

advisory—Advice and information provided to assist a pilot in the safe conduct of flight and aircraft movement.

aeronautical chart—A map used in air navigation.

1. Sectional Charts (1:500,000)—Designed for visual navigation of slow or medium speed aircraft. Topographic information on these charts features the portrayal of relief, and a selection of visual check points for VFR flight. Aeronautical information includes visual and radio aids to navigation, airports, controlled airspace, restricted areas, obstructions and related data.

2. VFR Terminal Area Charts (1:250,000)—Depict Terminal Control Area (TCA) airspace which provides for the control or segregation of all the aircraft within the TCA.

3. World Aeronautical Charts (WAC) (1:1,000,000)—Provide a standard series of aeronautical charts covering land areas of the world, at a size and scale convenient for navigation by moderate speed aircraft.

4. En Route Low Altitude Charts—Provide aeronautical information for en route instrument navigation (IFR) in the low altitude stratum. Area charts which are a part of this series furnish terminal data at a large scale in congested areas.

5. En Route High Altitude Charts—Provide aeronautical information for en route instrument navigation (IFR) in the high altitude stratum.

6. Area Navigation (RNAV) High Altitude Charts—Provide aeronautical information for en route IFR navigation for high altitude air routes established for aircraft equipped with RNAV systems.

7. Instrument Approach Procedures (IAP) Charts—Portray the aeronautical data which is required to execute an instrument approach to an airport.

8. Standard Instrument Departure (SID) Charts—Designed to expedite clearance delivery and to facilitate transition between take-off and en route operations. Each SID procedure is presented as a separate chart and may serve a single airport or more than one airport in a given geographical location.

9. Standard Terminal Arrival (STAR) Charts—Designated to expedite air traffic control arrival route procedures and to facilitate transition between en route and instrument approach operations. Each STAR procedure is presented as a separate chart and may serve a single airport or more than one airport in a given geographical location.

10. Airport Taxi Charts—Designed to expedite the efficient and safe flow of ground traffic at an airport. These charts are identified by the official airport name.

affirmative—Yes

aircraft approach category—A grouping of aircraft based on a speed of 1.3 V_{so} (at maximum certificated landing weight), or on maximum certificated landing weight. If an aircraft falls into two categories, it is placed in the higher of the two. The categories are as follows:

1. Category A—Speed less than 91 knots; weight less than 30,001 pounds.

2. Category B—Speed 91 knots or more but less than 121 knots; weight 30,001 pounds or more but less than 60,001 pounds.

3. Category C—Speed 121 knots or more but less than 141 knots; weight 60,001 pounds or more but less than 150,001 pounds.

4. Category D—Speed 141 knots or more but less than 166 knots; weight 150,001 pounds or more.

5. Category E—Speed 166 knots or more; any weight.

aircraft classes—For the purposes of Wake Turbulence Separation Minima, ATC classifies aircraft as follows:

1. Heavy—Aircraft capable of takeoff weights of 300,000 pounds or more whether or not they are operating at this weight during a particular phase of flight.

2. Large—Aircraft of more than 12,500 pounds, maximum certificated takeoff weight, up to 300,000 pounds.

3. Small—Aircraft of 12,500 pounds or less, maximum certificated takeoff weight.

Air Defense Identification Zone/ADIZ—The area of airspace over land or water, extending upward from the surface, within which the ready identification, the location, and the control of aircraft are required in the interest of national security.

1. Domestic Air Defense Identification Zone—An ADIZ within the United States.

2. Coastal Air Defense Identification Zone—An ADIZ over the coastal waters of the United States.

3. Distance Early Warning Identification Zone (DEWIZ)—An ADIZ over the coastal waters of the State of Alaska.

Airman's Information Manual/AIM—A publication designed primarily as a pilot's operational and instructional manual for use in the National Airspace System of the United States. It consists of the following basic parts which may be purchased separately from the Superintendent of Documents, Government Printing Office, Washington, D.C. 20402.

Basic Flight Information and ATC Procedures.

Airport/Facility Directory.

Notices to Airman.

Graphic Notices and Supplemental Data.

AIRMET/Airman's Meteorological Information—Inflight weather advisories which cover moderate icing, moderate turbulence, sustained winds of 30 knots or more within 2000 feet of the surface and the initial onset of phenomena producing extensive areas of visibilities below 3 miles or ceilings less than 1000 feet. It concerns weather phenomena which are of operational interest to all aircraft and potentially hazardous to aircraft having limited capability because of lack of equipment, instrumentation or pilot qualifications. It concerns weather of less severity than SIGMETs.

Airport Advisory Area—The area within five statute miles of an airport not served by a control tower (i.e., there is no tower or the tower is not in operation) on which is located a Flight Service Station.

Airport Advisory Service/AAS—A service provided by Flight Service Stations at airports not served by a control tower. This service consists of providing information to arriving and departing aircraft concerning wind direction and speed, favored runway, altimeter setting, pertinent known traffic, pertinent known field conditions, airport taxi routes and traffic patterns, and authorized instrument approach procedures. This information is advisory in nature and does not constitute an ATC clearance.

airport elevation/field elevation—The highest point of an airport's usable runways measured in feet from mean sea level.

airport lighting—Various lighting aids that may be installed on an airport. Types of airport lighting include:

1. Approach Light System/ALS—An airport lighting facility which provides visual guidance to landing aircraft by radiating light beams in a directional pattern by which the pilot aligns the aircraft with the extended centerline of the runway on his final approach for landing.

2. Runway lights/runway edge lights—Lights used to define the lateral limits of a runway. Runway lights are uniformly spaced at intervals of approximately 200 feet, and the intensity may be controlled or preset.

3. Touchdown zone lighting—Two rows of transverse light bars located symmetrically about the runway centerline normally at 100 foot intervals. The basic system extends 3,000 feet along the runway.

4. Runway centerline lighting—Flush centerline lights spaced at 50-foot intervals beginning 75 feet from the landing threshold and extending to within 75 feet of the opposite end of the runway.

5. Threshold lights—Fixed green lights arranged symmetrically left and right of the runway centerline, identifying the runway threshold.

6. Runway End Identifier Lights/REIL—Two synchronized flashing lights, one on each side of the runway threshold, which provide rapid and positive identification of the approach end of a particular runway.

7. Visual Approach Slope Indicator/VASI—An airport lighting facility providing vertical visual approach slope guidance to aircraft during approach to landing by radiating a directional pattern of high intensity red and white focused light beams which indicate to the pilot that he is "on path," if he sees red/white; "above path" if white/white; and "below path" if red/red. Some airports serving large aircrafts have three-bar VASIs which provide two visual glide paths to the same runway.

8. Boundary lights—Lights defining the perimeter of an airport or landing area.

airport rotating beacon—At civil airports, alternating white and green flashes indicate the location of the airport. The total number of flashes are 12 to 15 per minute. At military airports, the beacons flash alternatively white and green, but are differentiated from civil beacons by dual-peaked (two quick) white flashes between the green flashes. Normally, operation of an airport rotating beacon during the hours of daylight means that the reported ground visibility at the airport is less than three miles and/or the reported ceiling is less than 1000 feet and, therefore, an ATC clearance is required for landing or takeoff.

Airport Surveillance Radar/ASR—Approach control radar used to detect and display an aircraft's position in the terminal area. ASR provides range and azimuth information but does not provide elevation data. Coverage of the ASR can extend up to 60 miles.

Airport traffic area—Unless otherwise specifically designated, that airspace within a horizontal radius of five statute miles from the geographical center of any airport at which a control tower is operating, extending from the surface up to, but not including, an altitude of 3000 feet above the elevation of the airport. Unless otherwise authorized or required by ATC, no person may operate an aircraft within an airport traffic area except for the purpose of landing at, or taking off from, an airport within that area. The purpose of an airport traffic area is to require the pilot establish communications with tower while in the area; also pilots of reciprocating aircraft must slow to 156 knots or less (200 knots for turbine aircraft).

Air Route Surveillance Radar/ARSR—Air route traffic control center (ARTCC) radar, used primarily to detect and display an aircraft's position while en route between terminal areas.

Air Route Traffic Control Center/ARTCC—A facility established to provide air traffic control service to aircraft operating on IFR flight plans within controlled airspace and principally during the en route phase of flight. When equipment capabilities and controller workload permit, certain advisory/assistance services may be provided to VFR aircraft.

airspeed—The speed of an aircraft relative to its surrounding air mass. The unqualified term "airspeed" means one of the following:

1. Indicated airpseed—The speed shown on the aircraft airspeed indicator. This is the speed used in pilot/controller communications under the general term "airspeed."

2. True airspeed—The airspeed of an aircraft relative to undisturbed air. Used primarily in flight planning and en route portion of flight. When used in pilot/controller communications, it is referred to as "true airspeed" and not shortened to "airspeed."

air traffic—Aircraft operating in the air or on an airport surface, exclusive of loading ramps and parking areas.

air traffic clearance/ATC clearance/clearance—An authorization by air traffic control, for the purpose of preventing collision *between known aircraft*, for an aircraft to proceed under specified traffic conditions within controlled airspace.

Air Traffic Control/ATC—A service operated by appropriate authority to promote the safe, orderly and expeditious flow of air traffic.

altitude—The height of a level, point or object measured in feet *Above Ground Level (AGL)* or from *Mean Sea Level (MSL)*.

1. MSL altitude—Altitude, expressed in feet measured from mean sea level.

2. AGL altitude—Altitude expressed in feet measured above ground level.

3. Indicated altitude—The altitude as shown by an altimeter. On a pressure or barometric altimeter, it is altitude as shown uncorrected for instrument error and uncompensated for variation from standard atmospheric conditions.

altitude readout/automatic altitude report—An aircraft's altitude, transmitted via the Mode C transponder feature, that is visually displayed in 100 foot increments on a radar scope having readout capability.

altitude restriction—An altitude or altitudes stated in the order flown, which are to be maintained until reaching a specific point or time.

approach clearance—Authorization by ATC for a pilot to conduct an instrument approach. The type of instrument approach for which cleared and other pertinent information is provided in the approach clearance when required. (See *cleared for approach.*)

approach control service—Air traffic control service provided by an approach control facility for arriving and departing VFR/IFR aircraft and, on occasion, en route aircraft. At some airports not served by an approach control facility, the ARTCC provides limited approach control service.

approach gate—The point on the final approach course which is one mile from the final approach fix on the side away from the airport or five miles from landing threshold, whichever is farther from the landing threshold. This is an imaginary point used within ATC as a basis for final approach course interception for aircraft being vectored to the final approach course.

approach sequence—The order in which aircraft are positioned while on approach or awaiting approach clearance.

approach speed—The recommended speed contained in aircraft manuals used by pilots when making an approach to landing.

arc—The track over the ground of an aircraft flying at a constant distance from a navigational aid by reference to distance measuring equipment (DME).

Area Navigation/RNAV—A method of navigation that permits aircraft operations on any desired course within the coverage of station-reference navigation signals or within the limits of self-contained system capability.

1. Area Navigation Low Route—An area navigation route within the airspace extending upward from 1200 feet above the surface of the earth to, but not including 18,000 feet MSL.

2. Area Navigation High Route—An area navigation route within the airspace extending upward from and including 18,000 feet MSL to flight level 450.

3. Random Area Navigation Routes/Random RNAV Routes—Direct routes, based on area navigation capability, between waypoints defined in terms of degree/distance fixes or offset from published or established routes/airways at specified distance and direction.

4. RNAV Waypoint/W/P—A predetermined geographical position used for route or instrument approach definition or progress reporting purposes that is defined relative to a VORTAC station position.

ATC advises—Used to prefix a message of noncontrol information when it is relayed to an aircraft by other than an air traffic controller.

ATC clears—Used to prefix an ATC clearance when it is relayed to an aircraft by other than an air traffic controller.

ATC instruction—Directives issued by air traffic control for the purpose of requiring a pilot to take specific actions.

ATC requests—Used to prefix an ATC request when it is relayed to an aircraft by other than an air traffic controller.

Automated Radar Terminal Systems/ARTS—In general, an ARTS displays for the terminal controller aircraft identification, flight plan data, other flight associated information, e.g., altitude and speed, and aircraft position symbols in conjunction with his radar presentation. Normal radar coexists with the alphanumeric display. In addition to enhancing visualization of the air traffic situation, ARTS facilitate intra/interfacility transfer and coordination of flight information. These capabilities are enabled by specially designed computers and subsystems tailored to the radar and communications equipments and operational requirements of each automated facility.

automatic altitude reporting—That function of a transponder which responds to Mode C interrogations by transmitting the aircraft's altitude in 100-foot increments.

Automatic Direction Finder/ADF—An aircraft radio navigation system which senses and indicates the direction to a L/MF nondirectional radio beacon (NDB) ground transmitter. Direction is indicated to the pilot as a magnetic bearing or as a relative bearing to the longitudinal axis of the aircraft depending on the type of indicator installed in the aircraft.

Automatic Terminal Information Service/ATIS—The continuous broadcast of recorded noncontrol information in selected terminal areas. Its purpose is to improve controller effectiveness and to relieve frequency congestion by automating the repetitive transmission of essential but routine information.

bearing—The horizontal direction to or from any point, usually measured clockwise from true north, magnetic north or some other reference point, through 360 degrees.

below minimums—Weather conditions below the minimums prescribed by regulation for the particular action involved, e.g., landing minimums, takeoff minimums.

call-up—Initial voice contact between a facility and an aircraft, using the identification of the unit being called and the unit initiating the call.

cardinal altitudes or flight levels—"Odd" or "Even" thousand-foot altitudes or flight levels; e.g., 5000, 6000, 7000, FL 250, FL 260, FL 270.

ceiling—The heights above the earth's surface of the lowest layer of clouds or obscuring phenomena that is reported as *broken, overcast,* or *obscuration*, and not classified as *thin* or *partial*.

circle to land—A maneuver initiated by the pilot to align the aircraft with a runway for landing when a straight-in landing from an instrument approach is not possible or is not desirable. This maneuver is made only after ATC authorization has been obtained and the pilot has established required visual reference to the airport.

clear-air turbulence/CAT—Turbulence encountered in air where no clouds are present. This term is commonly applied to high-level turbulence associated with wind shear. CAT is often encountered in the vicinity of the jet stream.

clearance limit—The fix, point, or location to which an aircraft is cleared when issued an air traffic clearance.

clearance void if not off by (time)—Used by ATC to advise an aircraft that the departure clearance is automatically cancelled if takeoff is not made prior to a specified time. The pilot must obtain a new clearance or cancel his IFR flight plan if not off by the specified time.

cleared as filed—Means the aircraft is cleared to proceed in accordance with the route of flight filed in the flight plan. This clearance does not include the altitude, SID, or SID Transition.

cleared for (type of) approach—ATC authorization for an aircraft to execute a specific instrument approach procedure to an airport: e.g., "Cleared for ILS runway 36 approach."

cleared for approach—ATC authorization for an aircraft to execute any standard or special instrument approach procedure for that airport. Normally, an aircraft will be cleared for a specific instrument approach procedure. "Cleared for the approach" is a clearance to begin a descent to the applicable assigned altitude, MEA, MOCA, MSA, etc. Upon receipt of such a clearance, a pilot must report leaving an assigned altitude. It is a good practice to report the altitude to which the pilot intends to descend; e.g., departing niner thousand for five thousand.

cleared for takeoff—ATC authorization for an aircraft to depart. Takeoff and landing clearance grant *permission* to execute the specified maneuver. As such they are not instructive; a pilot may refuse a takeoff or landing clearance as in the case of possible wake turbulence.

cleared for the option—ATC authorization for an aircraft to make a touch-and-go, low approach, missed approach, stop and go, or full stop landing at the discretion of the pilot. It is normally used in training so that an instructor can evaluate a student's performance under changing situations.

cleared through—ATC authorization for an aircraft to make intermediate stops at specified airports without refiling a flight plan while en route to the clearance limit.

cleared to land—ATC authorization for an aircraft to land. It is predicated on known traffic and known physical airport conditions. (See *cleared for takeoff*).

clear of traffic—Previously issued traffic is no longer a factor.

climb to VFR—ATC authorization for an aircraft to climb to VFR conditions within a control zone when the only weather limitation is restricted visibility. The aircraft must remain clear of clouds while climbing to VFR.

Codes/transponder codes—The number assigned to a reply signal transmitted by a transponder.

combined station/tower/CS/T—An air traffic control facility which combines the functions of a flight service station and an airport traffic control tower.

compass locator—A low power, low or medium frequency (L/MF) radio beacon installed at the site of the outer or middle marker of an instrument landing system (ILS). It can be used for navigation at distances of approximately 15 miles or as authorized in the approach procedure.

    1. *Outer Compass Locator/LOM*—A compass locator installed at the site of the outer marker of an instrument landing system.

    2. *Middle Compass Locator/LMM*—A compass locator installed at the site of the middle marker of an instrument landing system.

compass rose—A circle graduated in degrees, printed on some charts or marked on the ground at an airport. It is used as a reference to either true or magnetic direction.

compulsory reporting points—Reporting points which must be reported to ATC. They are designated on aeronautical charts by solid triangles or filed in a flight plan as fixes selected to define direct routes. These points are geographical locations which are defined by navigation aids/fixes. Pilots should discontinue position reporting over compulsory reporting points when informed by ATC that their aircraft is in "radar contact." This does *not* imply the discontinuance of reporting leaving an assigned altitude.

contact—
1. Establish communications.
2. A flight condition wherein the pilot ascertains the altitude of his aircraft and navigates by visual reference to the surface.

contact approach—An approach wherein an aircraft on an IFR flight plan, operating clear of clouds with at least one mile flight visibility and having an air traffic control authorization, may deviate from the instrument approach procedure and proceed to the airport of destination by visual reference to the surface. This approach will only be authorized when requested by the pilot and the reported ground visibility at the destination is at least one statute mile.

controlled airspace—Airspace, designated as a continental control area, control area, control zone, terminal control area, or transition area, within which some or all aircraft may be subject to air traffic control.

Control sector—An airspace area of defined horizontal and vertical dimensions for which a controller, or group of controllers, has air traffic control responsibility, normally within an air route traffic control center or an approach control facility. Sectors are established based on predominant traffic flows, altitude strata, and controller workload. Pilot communications during operations within a sector normally maintained on discrete frequencies assigned to the sector.

course—
1. The intended direction of flight in the horizontal plane measured in degrees from north.
2. The ILS localizer signal pattern usually specified as front course or back course.

cross (fix) at (altitude)—Used by ATC when a specific altitude restriction at a specified fix is required.

critical engine—The engine which, upon failure, would most adversely affect the performance or handling qualities of an aircraft.

cross (fix) at or above (altitude)—Used by ATC when an altitude restriction at a specified fix is required. It does not prohibit the aircraft from crossing the fix at a higher altitude than specified; however, the higher altitude may not be one that will violate a succeeding altitude restriction or altitude assignment.

cross (fix at or below (altitude)—Used by ATC when a maximum crossing altitude at a specific fix is required. It does not prohibit the aircraft from crossing the fix at a lower altitude; however, it must be at or above the minimum IFR altitude.

crosswind—

1. When used concerning the traffic pattern, the word means "crosswind leg."

2. When used concerning wind conditions, the word means a wind not parallel to the runway or the path of an aircraft.

cruise—Used in an ATC clearance to authorize a pilot to conduct flight at any altitude from the minimum IFR altitude up to and including the altitude specified in the clearance. The pilot may level off at any intermediate altitude within this block of airspace. Climb/descent within the block is to be made at the discretion of the pilot. However, once the pilot starts descent and reports leaving an altitude in the block, he may not return to that altitude without additional ATC clearance. Further, it is approval for the pilot to proceed to and make an approach at destination airport and can be used in conjunction with:

1. An airport clearance limit at locations with a standard/special instrument approach procedure. The FARs require that if an instrument letdown to an airport is necessary, the pilot shall make the letdown in accordance with a standard/special instrument approach procedure for that airport, *or*

2. An airport clearance limit at locations that are within/below/outside controlled airspace and without a standard/special instrument approach procedure. Such a clearance is *not authorization* for the pilot to descend under IFR conditions below the applicable minimum IFR altitude nor does it imply that ATC is exercising control over aircraft in uncontrolled airspace; however, it provides a means for the aircraft to proceed to destination airport, descend and land in accordance with applicable FARs governing VFR flight operations. Also, this provides search and rescue protection until such time as the IFR flight plan is closed.

decision height/DH—With respect to the operation of aircraft means, the height at which a decision must be made during an ILS or PAR instrument approach, to either continue the approach or to execute a missed approach.

departure control—A function of an approach control facility providing air traffic control service for departing IFR and, under certain conditions, VFR aircraft.

departure time—The time an aircraft becomes airborne.

deviations—

1. A departure from a current clearance; such as an off course maneuver to avoid weather or turbulence.

2. Where specifically authorized in the FARs and requested by the pilot ATC may permit pilots to deviate from certain regulations.

DF approach procedure—Used under emergency conditions where another instrument approach procedure cannot be executed. DF guidance for an instrument approach is given by ATC facilities with DF capability.

DF fix—The geographical location of an aircraft obtained by one or more direction finders.

DF guidance/DF Steer—Headings provided to aircraft by facilities equipped with direction finding equipment. These headings, if followed, will lead the aircraft to a predetermined point such as the DF station or an airport. DF guidance is given to aircraft in distress or to other aircraft which request the service. Practice DF guidance is provided when workload permits.

direct—Straight line flight between two navigational aids, fixes, points or any combination thereof. When used by pilots in describing off-airway routes, points defining direct route segments become compulsory reporting points unless the aircraft is under radar contact.

direction finder—A radio receiver equipped with a directional sensing antenna used to take bearings on a radio transmitter.

discrete code—As used in the Air Traffic Control Radar Beacon System (ATCRBS), any one of the 4096 selectable Mode 3/A aircraft transponder codes except those ending in zero zero; e.g., discrete codes: 0010, 1201, 2317, 7777; non-discrete codes: 0100, 1200, 7700. Non-discrete codes are normally reserved for radar facilities that are not equipped with discrete decoding capability and for other purposes such as emergencies (7700), VFR aircraft (1200), etc.

displaced threshold—A threshold that is located at a point on the runway other than the designated beginning of the runway.

Distance Measuring Equipment/DME—Equipment (airborne and ground) used to measure, in nautical miles, the slant range distance of an aircraft from the DME navigational aid.

DME fix—A geographical position determined by reference to a navigational aid which provides distance and azimuth information. It is defined by a specific distance in nautical miles and a radial or course (i.e., localizer) in degrees magnetic from that aid.

DME separation—Spacing of aircraft in terms of distances (nautical miles) determined by reference to distance measuring equipment (DME).

Emergency Locator Transmitter/ELT—A radio transmitter attached to the aircraft structure which operates from its own power source on 121.5 MHz and 243.0 MHz. It aids in locating downed aircraft by radiating a downward sweeping audio tone, 2-4 times per second. It is designed to function without human action after an accident.

en route air traffic control services—Air traffic control service provided aircraft on an IFR flight plan, generally by centers, when these aircraft are operating between departure and destination terminal areas. When equipment capabilities and controller workload permit, certain advisory/assistance services may be provided to VFR aircraft.

en route flight advisory service/Flight Watch—A service specifically designed to provide, upon pilot request, timely weather information pertinent to his type of flight, intended route of flight and altitude. The FSSs providing this service are listed in the AIM.

execute missed approach—Instructions issued to a pilot making an instrument approach which means continue inbound to the missed approach point and execute the missed approach procedure as described on the Instrument Approach Procedure Chart, or as previously assigned by ATC. The pilot may climb immediately to the altitude specified in the missed approach procedure upon making a missed approach. No turns should be initiated prior to reaching the missed approach point. When conducting an ASR or PAR approach, execute the assigned missed approach procedure immediately upon receiving instructions to "execute missed approach."

expect approach clearance (time)/EAC—The time at which it is expected that an arriving aircraft will be cleared to commence an approach for landing. It is issued when the aircraft clearance limit is a designated Initial, Intermediate, or Final Approach Fix for the approach in use and the aircraft is to be held. If delay is anticipated, the pilot should be advised of his EAC at least five minutes before the aircraft, is estimated to reach the clearance limit.

expect (altitude) at (time) or (fix)—Used under certain conditions in a departure clearance to provide a pilot with an altitude to be used in the event of two-way communications failure.

expect further clearance (time)/EFC—The time at which it is expected that additional clearance will be issued to an aircraft. It is issued when the aircraft clearance limit is a fix not designated as part of the approach procedure to be executed and the aircraft will be held. If delay is anticipated, the pilot should be advised of his EFC at least five minutes before the aircraft is estimated to reach the clearance limit.

feeder route—A route depicted on instrument approach procedure charts to designate routes for aircraft to proceed from the en route structure to the initial approach fix (IAF).

final—Commonly used to mean that an aircraft is on the final approach course or is aligned with a landing area.

final approach course—A straight line extension of a localizer, a final approach radial/bearing, or a runway centerline, all without regard to distance.

final approach fix/FAF—The designated fix from or over which the final approach (IFR) to an airport is executed. The FAF identifies the beginning of the final approach segment of the instrument approach.

final approach-IFR—The flight path of an aircraft which is inbound to an airport on a final instrument approach course, beginning at the final approach fix or point and extending to the airport or the point where a circle to land maneuver or a missed approach is executed.

fix—A geographical position determined by visual reference to the surface, by reference to one or more radio NAVAIDs, or by another navigational device.

flag—A warning device incorporated in certain airborne navigation and flight instruments indicating that:

1. Instruments are inoperative or otherwise not operating satisfactorily, or

2. Signal strength or quality of the received signal falls below acceptable values.

flight level—A level of constant atmospheric pressure related to a reference datum of 29.92 inches of mercury. Each is stated in three digits that represent hundreds of feet. For example, flight level 250 represents a barometric altimeter indication of 25,000 feet; flight level 255, an indication of 25,500 feet.

flight plan—Specified information relating to the intended flight of an aircraft that is filed orally or in writing with an FSS or an ATC facility.

Flight Service Station/FSS—Air Traffic Service facilities within the National Airspace System (NAS) which provide pilot briefing and en route communications with VFR flights, assist lost IFR/VFR aircraft, assist aircraft having emergencies, relay ATC clearances, originate, classify, and disseminate Notices to Airmen, broadcast aviation weather and NAS information, receive and close flight plans, monitor radio NAVAIDS, notify search and rescue units of missing VFR aircraft, and operate the national weather tele-typewriter systems. In addition, at selected locations, FSSs take weather observations, issue airport advisories, administer airmen written examinations, and advise Customs and Immigration of transborder flight.

flight time—The time from the moment the aircraft first moves under its own power for the purpose of flight until the moment it comes to rest at the next point of landing. ("Block-to-block" time.)

266

Flight Watch—A shortened term for use in air-ground contacts on frequency 122.0 MHz to identify the flight service station providing En Route Flight Advisory Service; e.g., "Oakland Flight Watch."

fly heading (degrees)—Informs the pilot of the heading he should fly. The pilot may have to turn to, or continue on, a specific compass direction in order to comply with the instructions. The pilot is expected to turn in the shorter direction to the heading, unless otherwise instructed by ATC.

glide path (on/above/below)—Used by ATC to inform an aircraft making a PAR approach of its vertical position (elevation) relative to the descent profile. The terms "slightly" and "well" are used to describe the degree of deviation; e.g., "slightly above glide path." Trend information is also issued with respect to the elevation of the aircraft and may be modified by the terms "rapidly" and "slowly", e.g., "well above glidepath, coming down rapidly."

glide slope/gs—Provides vertical guidance for aircraft during approach and landing. The glide slope consists of the following:

1. Electronic components emitting signals which provide vertical guidance by reference to airborne instrument during instrument approaches such as ILS; *or*

2. Visual ground aids such as VASI which provides vertical guidance for VFR approach or for the visual portion of an instrument approach and landing.

glide slope intercept altitude—The minimum altitude of the intermediate approach segment prescribed for a precision approach which assures required obstacle clearance.

go around—Instructions for a pilot to abandon his approach to landing. Additional instructions may follow. Unless otherwise advised by ATC, a VFR aircraft or an aircraft conducting a visual approach should overfly the runway while climbing to traffic pattern altitude and enter the traffic pattern via the crosswind leg. A pilot on an IFR flight plan making an instrument approach should execute the published missed approach procedure or process as instructed by ATC; e.g., "Go Around" (additional instructions, if required).

Ground Controlled Approach/GCA—A radar approach system operated from the ground by air traffic control personnel transmitting instructions to the pilot by radio. The approach may be conducted with surveillance radar (ASR) only or with both surveillance and precision approach radar (PAR). Usage of the term "GCA" by pilots is discouraged except when referring to a GCA facility. Pilots should specifically request a "PAR" approach when a precision radar approach is desired or request an "ASR" or "surveillance" approach when a nonprecision radar approach is desired. For the most part, GCA's are miliary facilities and are not available to general aviation pilots except in an emergency.

ground speed—The speed of an aircraft relative to the surface of the earth.

handoff—Transfer of radar identification of an aircraft from one controller to another, either within the same facility or interfacility. Actual transfer to control responsibility may occur at the time of the handoff, or at a specified time, point or altitude.

have numbers—Used by pilots to inform ATC that they have received runway and wind information only.

height above airport/HAA—The height of the Minimum Descent Altitude above the published airport elevation. This is published in conjunction with circling minimums.

height above touchdown/HAT—The height of the Decision Height or Minimum Descent Altitude above the highest runway elevation in the touchdown zone (first 3,000 feet of the runway). HAT is published on instrument approach charts in conjunction with all straight-in minimums.

hold—A predetermined maneuver which keeps aircraft within a specified airspace while awaiting further clearance from air traffic control. Also used during ground operations to keep aircraft within a specified area or at a specified point while awaiting further clearance from air traffic control.

holding fix—A specified fix identifiable to a pilot by NAVAIDS or visual reference to the ground used as a reference point in establishing and maintaining the position of an aircraft while holding.

homing—Flight toward a NAVAID, without correcting for wind, by adjusting the aircraft heading to maintain a relative bearing of zero degrees.

ident—A request for a pilot to activate the aircraft transponder identification feature. This will help the controller to conform an aircraft identity or to identify an aircraft.

if no transmission received for (time)—Used by ATC in radar approaches to prefix procedures which should be followed by the pilot in event of lost communications.

IFR conditions—Weather conditions below the minimum for flight under visual flight rules.

IFR over the top—The operation of an aircraft over the top on an IFR flight plan when cleared by air traffic control to maintain "VFR conditions" or "VFR conditions on top."

ILS categories—

    1. ILS Category I—An ILS approach procedure which provides for approach to a height above touchdown of not less than 200 feet and with runway visual range of not less than 1800 feet.

2. ILS Category II—An ILS approach procedure which provides for approach to a height above touchdown of not less than 100 feet and with runway visual range of not less than 1200 feet.

3. ILS Category III—

a. IIIA—An ILS approach procedure which provides for approach without a decision height minimum and with runway visual range of not less than 700 feet.

b. IIIB—An ILS approach procedure which provides for approach without a decision height minimum and with runway visual range of not less than 150 feet.

c. IIIC—An ILS approach procedure which provides for approach without a decision height minimum and without runway visual range minimum.

immediately—Used by ATC when such action compliance is required to avoid an imminent situation.

initial approach Fix/IAF—The fixes depicted on instrument approach procedure charts that identifies the beginning of the initial approach segment(s).

inner marker/IM—A marker beacon used with an ILS (CAT II) precision approach located between the middle marker and the end of the ILS runway, transmitting a radiation pattern keyed at six dots per second and indicating to the pilot, both aurally and visually, that he is at the designated decision height (DH), normally 100 feet above the touchdown zone elevation, on the ILS CAT II approach. It also marks progress during a CAT III approach.

Instrument Flight Rules/IFR—Rules governing the procedures for conducting instrument flight. Also a term used by pilots and controllers to indicate type of flight plan.

Instrument Landing System/ILS—A precision instrument approach system consisting of the following electronic components and visual aids:

1. Localizer (See *localizer*).
2. Glide Slope (See *glide slope*).
3. Outer Marker (See *outer marker*).
4. Middle Marker (See *middle marker*).
5. Approach Lights.

instrument runway—A runway equipped with electronic and visual navigation aids for which precision or nonprecision approach procedure having straight-in landing minimums has been approved.

intermediate fix/IF—The fix that identifies the beginning of the intermediate approach segment of an instrument approach procedure. The fix is not normally identified on the instrument approach chart as an intermediate fix (IF).

international airport—Relating to international flight, it means:

1. An airport of entry which has been designated by the Secretary of Treasury or Commissioner of Customs as an internal airport for customs service.

2. A landing rights airport at which specific permission to land must be obtained from customs authorities in advance of contemplated use.

3. Airports designated under the Convention on International Civil Aviation as an airport for use by international commercial air transport and/or international general aviation.

intersection—

1. A point defined by any combination of courses, radials, or bearings of two or more navigational aids.

2. Used to describe the point where two runways cross, a taxiway and a runway cross or, two taxiways cross.

intersection departure/intersection takeoff—A takeoff or proposed takeoff on a runway from an intersection.

I say again—The message will be repeated.

jet route—A route designed to serve aircraft operations from 18,000 feet MSL up to and including flight level 450. The routes are referred to as "J" routes with numbering to identify the designated route; e.g., J 105.

known traffic—With respect to ATC clearances, means aircraft whose altitude, position and intentions are known to ATC.

landing minimums (IFR)—Descent below the established MDA or DH is not authorized during an approach unless the aircraft is in a position from which a normal approach to the runway of intended landing can be made, and adequate visual reference to required visual cues is maintained.

landing sequence—The order in which aircraft are positioned for landing.

last assigned altitude—The last altitude/flight level assigned by ATC *and acknowledged* by the pilot.

Limited Remote Communications Outlet/LRCO—An unmanned satellite air/ground communications facility which may be associated with a VOR. These outlets effectively extend the service range of the FSS and provide greater communications reliability. LRCO's are depicted on En Route Charts.

localizer—The component of an ILS which provides course guidance to the runway.

localizer type directional aid/LDA—A NAVAID used for nonprecision instrument approaches with utility and accuracy comparable to a localizer but which is not a part of a complete ILS and is not aligned with the runway.

low altitude airway structure/federal airways—The network of airways serving aircraft operations up to but not including 18,000 feet MSL.

low altitude alert, check your altitude immediately—(See *safety advisory*).

low approach—An approach over an airport or runway following an instrument approach or a VFR approach including the go-around maneuver where the pilot intentionally does not make contact with the runway.

Mach number—The ratio of true airspeed to the speed of sound, e.g., Mach .82, Mach 1.6.

maintain—

1. Concerning altitude/flight level, the term means to remain at the altitude/flight level specified. The phrase "climb and" or "descend and" normally precede "maintain" and the altitude assignment; e.g., "descend and maintain 5000." If a SID procedure is assigned in the initial or subsequent clearance, the altitude restrictions in the SID, if any, will apply unless otherwise advised by ATC.

2. Concerning other ATC instructions, the term is used in its literal sense; e.g., maintain VFR.

make short approach—Used by ATC to inform a pilot to alter his traffic pattern so as to make a short final approach.

marker beacon—An electronic navigation facility transmitting a 75 MHz vertical fan or boneshaped radiation pattern. Marker beacons are identified by their modulation frequency and keying code and when received by compatible airborne equipment, indicate to the pilot, both aurally and visually, that he is passing over the facility.

Maximum Authorized Altitude/MAA—A published altitude representing the maximum usable altitude or flight level for an airspace structure or route segment. It is the highest altitude on a Federal airway, Jet route, area navigation low or high route, or other direct route for which an MEA is designated in FAR Part 95, at which adequate reception of navigation and signals is assured.

Mayday—The international radiotelephony distress signal. When repeated three times, it indicates imminent and grave danger and that immediate assistance is requested.

Microwave Landing System/MLS—An instrument landing system operating in the microwave spectrum which provides lateral and vertical guidance to aircraft having compatible avionics equipment.

middle marker/MM—A marker beacon that defines a point along the glide slope of an ILS normally located at or near the point of decision height (ILS Category I). It is keyed to transmit alternate dots and

271

dashes, two per second, on a 1300 Hz tone which is received aurally and visually by compatible airborne equipment.

Military Operations Area/MOA—An MOA is an airspace assignment of defined vertical and lateral dimensions established outside positive control area to separate/segregate certain military activities from IFR traffic and to identify for VFR traffic where these activities are conducted.

Minimum Crossing Altitude/MCA—The lowest altitude at certain fixes at which an aircraft must cross when proceeding in the direction of a higher minimum en route IFR altitude (MEA).

Minimum Descent Altitude/MDA—The lowest altitude, expressed in feet above mean sea level, to which descent is authorized on final approach or during circle-to-land maneuvering in execution of a standard instrument approach procedure where no electronic glide slope is provided.

Minimum En Route IFR Altitude/MEA—The lowest published altitude between radio fixes which assures acceptable navigational signal coverage and meets obstacle clearance requirements between those fixes. The MEA prescribed for a Federal airway or segment thereof, area navigational low or high route, or other direct route, applies to the entire width of the airway, segment or route between the radio fixes defining the airway, segment or route.

Minimum Holding Altitude/MHA—The lowest altitude prescribed for a holding pattern which assures navigational signal coverage, communications, and meets obstacle clearance requirements.

minimum IFR altitudes—Minimum altitudes for IFR operations as prescribed in FAR Part 91. These altitudes are published on aeronautical charts and prescribed in FAR Part 95, for airways and routes, and FAR Part 97, for standard instrument approach procedures. If no applicable minimum altitude is prescribed in FAR Parts 95 or 97, the following minimum IFR altitude applies:

1. In *designated* mountainous areas, 2000 feet above the highest obstacle within a horizontal distance of five statute miles from the course to be flown, or

2. Other than mountainous areas, 1000 feet above the highest obstacle within a horizontal distance of five statute miles from the course to be flown; or

3. As otherwise authorized by the Administrator or assigned by ATC.

Minimum Obstruction Clearance Altitude/MOCA—The lowest published altitude in effect between radio fixes on VOR airways, off-airway routes, or route segments which meets obstacle clearance requirements for the entire route segment and which assures acceptable navigation signal coverage only within 22 nautical miles of a VOR.

Minimum Reception Altitude/MRA—The lowest altitude at which an intersection can be determined.

Minimum Safe Altitude/MSA—
1. The minimum altitude specified in FAR Part 91, for various aircraft operations.
2. Altitudes depicted on instrument approach charts and identified as minimum sector altitudes or emergency safe altitudes which provide a minimum of 1000 feet obstacle clearance within a specified distance from the navigation facility upon which an instrument approach procedure is predicated. These altitudes are for *emergency use only* and do not necessarily guarantee NAVAID reception. Minimum sector altitudes are established for all procedures (except localizers without an NDB) within a 25 nautical mile radius of the navigation facility. Emergency safe altitudes are established for some military procedures within a 100 nautical mile radius of the navigation facility.

Minimum Safe Altitude Warning (MSAW)—A function of the ARTS III computer that aids the controller by alerting him when a tracked Mode C equipped aircraft is below or is predicted by the computer to go below a predetermined minimum safe altitude.

minimums/minima—Weather conditions requirements established for a particular operation or type of operation; e.g., IFR takeoff or landing, alternate airport for IFR flight plans, VFR flight.

Minimum Vectoring Altitude/MVA—The lowest MSL altitude at which an IFR aircraft will be vectored by a radar controller, except as otherwise authorized for radar approaches, departures and missed approaches. The altitude meets IFR obstacle clearance criteria. It may be lower than the published MEA along an airway or J-route segment. It may be utilized for radar vectoring only upon the controllers' determination that an adequate radar return is being received from the aircraft being controlled. Charts depicting minimum vectoring altitudes are normally available only to the controllers and not to pilots. If a pilot is below MVA, controller headings are "suggestions" only.

missed approach—
1. A maneuver conducted by a pilot when an instrument approach cannot be completed to a landing. The route of flight and altitude are shown on instrument approach procedure charts. A pilot executing a missed approach prior to the Missed Approach Point (MAP) must continue along the final approach to the MAP. The pilot may climb immediately to the altitude specified in the missed approach procedure.
2. A term used by the pilot to inform ATC that he is executing the missed approach.
3. At locations where ATC radar service is provided the pilot should

conform to radar vectors, when provided by ATC, in lieu of the published missed approach procedure.

Missed Approach Point/MAP—A point prescribed in each instrument approach procedure at which a missed approach procedure shall be executed if the required visual reference does not exist.

mode—The letter or number assigned to a specific pulse spacing of radio signals transmitted or received by ground interrogator or airborne transponder components of the Air Traffic Control Radar Beacon System (ATCRBS).

NAVAID classes—VOR, VORTAC, and TACAN aids are classed according to their operational use. The three classes of NAVAIDS are:
T - Terminal.
L - Low Altitude.
H - High Altitude.

negative—"No" or "Permission not granted" or "That is not correct."

negative contact—Used by pilots to inform ATC that:
1. Previously issued traffic is not in sight. It may be followed by the pilot's request for controller to provide assistance in avoiding the traffic.
2. They were unable to contact ATC on a particular frequency.

night—The time between the end of evening civil twilight and the beginning of morning civil twilight, as published in the American Air Almanac, converted to local time.

NOS—National Ocean Survey, National Oceanic and Atmospheric Administration. Publisher of VFR and IFR charts.

no gyro approach/vector—A radar approach/vector provided in case of a malfunctioning gyrocompass or directional gyro. Instead of providing the pilot with headings to be flown, the controller observes the radar track and issues control instructions "turn right/left" or "stop turn" as appropriate.

nondirectional beacon/NDB—A radio beacon transmitting nondirectional signals whereby the pilot of an aircraft equipped with direction finding equipment can determine his bearing to or from the radio beacon and "home" on or track to or from the station. When the radio beacon is installed in conjunction with the Instrument Landing System marker, it is normally called a *compass locator.*

nonprecision approach—A standard instrument approach procedure in which no electronic glide slope is provided; e.g., VOR, TACAN, NDB, LOC, ASR, LDA, or SDF approaches.

numerous targets vicinity—A traffic advisory issued by ATC to advise pilots that targets on the radar scope are too numerous to issue individually.

off-route vector—A vector by ATC which takes an aircraft off a previously assigned route. Altitudes assigned by ATC during such vectors provide required obstacle clearance.

olive branch routes—Training routes used by USAF and USN jet aircraft in both VFR and IFR weather conditions from the surface to the published altitude. Routes, their altitudes and times of operation are shown in the AIM Graphic Notices Volume. The current operational status of a particular route may be obtained by calling an FSS near the route.

on course—

1. Used to indicate that an aircraft is established on the route centerline.

2. Used by ATC to advise a pilot making a radar approach that his aircraft is lined up on the final approach course.

option approach—An approach requested and conducted by a pilot which will result in either a touch-and-go, missed approach, low approach, stop-and-go or full stop landing.

out—The conversion is ended and no response is expected.

outer fix—A general term used within ATC to describe fixes in the terminal area, other than the final approach fix. Aircraft are normally cleared to these fixes by an Air Route Traffic Control Center of an Approach Control Facility. Aircraft are normally cleared from these fixes to the final approach fix or final approach course.

outer marker/OM—A marker beacon at or near the glide slope intercept altitude of an ILS approach. It is keyed to transmit two dashes per second on a 400 Hz tone which is received aurally and visually by compatible airborne equipment. The OM is normally located four to seven miles from the runway threshold on the extended centerline of the runway.

over—My transmission is ended; I expect a response.

pan—The international radio-telephony urgency signal. When repeated three times indicates uncertainty or alert, followed by nature of urgency. (See *Mayday*).

PAR approach—A precision instrument approach wherein the air traffic controller issues guidance instructions, for pilot compliance, based on the aircraft's position in relation to the final approach course (azimuth), the glide slope (elevation), and the distance (range) from the touchdown point on the runway as displayed on the controller's radar scope.

pilot-in-command—The pilot responsible for the operation and safety of an aircraft during flight time.

Pilot's Automatic Telephone Weather Answering Service/PATWAS—A continuous telephone recording containing current and forecast weather information for pilots.

pilot's discretion—When used in conjunction with altitude assignments, means that ATC has offered the pilot the option of starting climb or descent whenever he wishes and conducting the climb or descent at any rate he wishes. He may temporarily level off at any intermediate altitude. However, once he has vacated an altitude he may not return to that altitude.

pilot weather report/PIREP—A report of meteorological phenomena encountered by aircraft in flight.

pilotage—Navigation by visual reference to landmarks.

position report/progress report—A report over a known location as transmitted by an aircraft to ATC.

positive control—The separation of all air traffic, within designated airspace, by air traffic control.

Positive Control Area/PCA—Airspace designated in FAR Part 71 wherein aircraft are required to be operated under Instrument Flight Rules (IFR). Vertical extent of PCA is from 18,000 feet to and including flight level 600 throughout most of the conterminous United States and from 18,000 feet to and including flight level 600 in designated portions of Alaska.

precision approach procedure—A standard instrument approach procedure in which an electronic glide slope is provided; e.g., ILS and PAR.

Precision Approach Radar/PAR—Radar equipment in some ATC facilities serving military airports, which is used to detect and display the azimuth, range, and elevation of an aircraft on the final approach course to a runway. It is used by air traffic controllers to provide the pilot with a precision approach, or to monitor certain nonradar approaches.

procedure turn—The maneuver prescribed when it is necessary to reverse direction to establish an aircraft on the intermediate approach segment or final approach course. The outbound course, direction of turn, distance within which the turn must be completed, and minimum altitude are specified in the procedure. However, the point at which the turn may be commenced, and the type and rate of turn, are left to the discretion of the pilot.

procedure turn inbound—That point of a procedure turn maneuver where course reversal has been completed and an aircraft is established inbound on the intermediate approach segment or final approach course. A report of "procedure turn inbound" is normally used by ATC as a position report for separation purposes.

profile descent—An uninterrupted descent (except where level flight is required for speed adjustment; e.g., 250 knots at 10,000 feet MSL) from cruising altitude/level to interception of a glide slope or to a minimum altitude specified for the initial or intermediate approach segment of a non-precision instrument approach. The profile descent normally terminates at the approach gate or where the glide slope or other appropriate minimum altitude is intercepted.

Prohibited Area—Designated airspace within which the flight of aircraft is prohibited.

published route—A route for which an IFR altitude has been established and published; e.g., Federal Airways, Jet Routes, Area Navigation Routes, Specified Direct Routes.

radar—A device which by measuring the time interval between transmission and reception of radio pulses and correlation the angular orientation of the radiated antenna beam or beams in azimuth and/or elevation, provides information on range, azimuth and/or elevation of objects in the path of the transmitted pulses.

    1. *Primary Radar*—A radar system in which a minute portion of a radio pulse transmitted from a site is reflected by an object and then received back at that site, for processing and display at an air traffic control facility.

    2. *Secondary Radar/Radar Beacon/ATCRBS*—A radar system in which the object to be detected is fitted with cooperative equipment in the form of a radio receiver/transmitter (transponder). Radar pulses transmitted from the searching transmitter/receiver (interrogator) site are received in the cooperative equipment and used to trigger a distinctive transmission from the transponder. This reply transmission rather than a reflected signal, is then received back at the transmitter/receiver site for processing and display at an air traffic control facility.

radar advisory—The provision of advice and information based on radar observations.

radar approach—An instrument approach procedure which utilizes Precision Approach Radar (PAR) or Airport Surveillance Radar (ASR).

radar contact—

    1. Used by ATC to inform an aircraft that it is identified on the radar display and radar service may be provided until radar identification is lost or radar service is terminated. When a pilot is informed of "radar contact," he automatically discontinues reporting over compulsory reporting points. The term *radar contact does not* imply that all headings given will assure adequate terrain clearance. If below Minimum Vectoring Altitude (MVA), headings are "suggestions" only.

    2. The term an air traffic controller uses to inform the transferring controller that the target being transferred is identified on his radar display.

radar contact lost—Used by ATC to inform a pilot that radar identification of his aircraft has been lost. The loss may be attributed to several things including the aircraft merging with weather or ground clutter, the aircraft flying below radar line of sight, the aircraft entering an area of poor radar return or a failure of the aircraft transponder or ground radar equipment.

radar environment—An area in which radar service may be provided.

radar flight following—The observation of the progress of radar identified aircraft, whose primary navigation is being provided by the pilot, wherein the controller retains and correlates the aircraft identity with the appropriate target or target symbol displayed on the radar scope.

radar identification—The process of ascertaining that an observed radar target is the radar return from a particular aircraft.

radar identified aircraft—An aircraft, the position of which has been correlated with an observed target or symbol on the radar display.

radar route —A flight path or route over which an aircraft is vectored. Navigational guidance and altitude assignments are provided by ATC.

radar service—A term which encompasses one or more of the following services based on the use of radar which can be provided by a controller to a pilot of a radar identified aircraft.

1. *Radar Separation*—Radar spacing of aircraft in accordance with established minima.

2. Radar Navigational Guidance—Vectoring aircraft to provide course guidance.

3. Radar Monitoring—The radar flight following of aircraft, whose primary navigation is being performed by the pilot, to observe and note deviations from its authorized flight path, airway, or route. When being applied specifically to radar monitoring of instrument approaches, i.e., with precision approach radar (PAR) or radar monitoring of simultaneous ILS approaches, it includes advice and instructions whenever an aircraft nears or exceeds the prescribed PAR safety limit or simultaneous ILS no transgression zone.

radar service terminated—Used by ATC to inform a pilot that he will no longer be provided any of the services that could be received while under radar contact. Radar service is automatically terminated and the pilot is not advised in the following cases:

1. An aircraft cancels its IFR flight plan, except within a TCA, TRSA, or where Stage II service is provided.

2. At the completion of a radar approach.

3. When an arriving aircraft receiving Stage I, II or III service is advised to contact the tower.

4. When an aircraft conducting a visual approach or contact approach is advised to contact the tower.

5. When an aircraft making an instrument approach has landed or the tower had the aircraft in sight, whichever occurs first.

radial—A magnetic bearing extending from a VOR/VORTAC/TACAN navigation facility.

radio altimeter/radar altimeter—Aircraft equipment which makes use of the reflection of radio waves from the ground to determine the height of the aircraft above the surface.

release time—A departure time restriction issued to a pilot by ATC when necessary to separate a departing aircraft from other traffic.

remote communications air/ground facility/RCAG—An unmanned VHF/UHF transmitter/receiver facility which is used to expand ARTCC air/ground communications coverage and to facilitate direct contact between pilots and controllers. RCAG facilities are sometimes not equipped with emergency frequencies 121.5 MHz and 243.0 MHz.

Remote Communications Outlet/RCO—An unmanned air/ground communications station remotely controlled, providing UHF and VHF transmit and receive capability to extend the service range of the FSS.

report—Used to instruct pilots to advise ATC of specified information, e.g., "Report passing Hamilton VOR."

reporting point—A geographical location in relation to which the position of an aircraft is reported.

request full route clearance/FRC—Used by pilots to request that the entire route of flight be read verbatim in an ATC clearance. Such request should be made to preclude receiving an ATC clearance based on the original filed flight plan when a filed IFR flight plan has been revised by the pilot, company, or operations prior to departure.

Restricted Area—Airspace designated under FAR Part 73 within which the flight of aircraft, while not wholly prohibited, is subject to restriction. Most restricted areas are designated joint use and IFR/VFR operations in the areas may be authorized by the controlling ATC facility when it is not being utilized by the using agency. Restricted areas are depicted on en route charts. Where joint use is authorized the name of the ATC controlling facility is also shown.

resume own navigation—Used by ATC to advise a pilot to resume his own navigational responsibility. It is issued after completion of a radar vector or when radar contact is lost while the aircraft is being radar vectored.

RNAV approach—An instrument approach procedure which relies on aircraft area navigation equipment for navigational guidance.

roger—I have received all of your last transmission. It should *not* be used to answer a question requiring a yes or no answer.

safety advisory—A safety advisory issued by ATC to aircraft under their control if ATC is aware the aircraft is at an altitude which, in the controller's judgment, places the aircraft in unsafe proximity to terrain, obstructions, or other aircraft. The controller may discontinue the issuance of further advisories if the pilot advises he is taking action to correct the situation or has the other aircraft in sight.

1. *Terrain/Obstruction Advisory*—A safety advisory issued by ATC to aircraft under their control if ATC is aware the aircraft is at an altitude which, in the controller's judgment, places the aircraft in unsafe proximity to terrain/obstructions; e.g., "Low Altitude Alert, check your altitude immediately."

2. *Aircraft Conflict Advisory*—A safety advisory issued by ATC to aircraft under their control if ATC is aware of an aircraft which is not under their control at an altitude which, in the controller's judgment, places both aircraft in unsafe proximity to each other. With the alert, ATC will offer the pilot an alternate course of action when feasible, e.g., "Traffic Alert, advise you turn right heading zero niner zero or climb to eight thousand immediately."

The issuance of a safety advisory is contingent upon the capability of the controller to have an awareness of an unsafe condition. The course of action provided will be predicated on other traffic under ATC control. Once the advisory is issued, it is solely the pilot's prerogative to determine what course of action, if any, he will take.

say again—Used to request a repeat of the last transmission. Usually specifies transmission or portion thereof not understood or received, e.g., "Say again all after ABRAM VOR."

say altitude—Used by ATC to ascertain an aircraft's specific altitude/ flight level. When the aircraft is climbing or descending, the pilot should state the indicated altitude rounded to the nearest 100 feet.

say heading—Used by ATC to request an aircraft heading. The pilot should state the actual heading of the aircraft.

segments of an instrument approach procedure—An instrument approach procedure may have as many as four separate segments depending on how the approach procedure is structured.

1. *Initial Approach*—The segment between the initial approach fix and the intermediate fix or the point where the aircraft is established on the intermediate course or final approach course.

2. *Intermediate Approach*—The segment between the intermediate fix or point and the final approach fix.

3. *Final Approach*—The segment between the final approach fix or point and the runway, airport or missed approach point.

4. *Missed Approach*—The segment between the missed approach point or point of arrival at decision height, and the missed approach fix at the prescribed altitude.

short range clearance—A clearance issued to a departing IFR flight which authorizes IFR flight to a specific fix short of the destination while air traffic control facilities are coordinating and obtaining the complete clearance.

sidestep maneuver—A visual maneuver accomplished by a pilot at the completion of an instrument approach to permit a straight-in landing on a parallel runway not more than 1200 feet to either side of the runway to which the instrument approach was conducted.

SIGMET/Significant Meteorological Information—A weather advisory issued concerning weather significant to the safety of all aircraft. SIGMET advisories cover tornados, lines of thunderstorms, embedded thunderstorms, large hail, severe and extreme turbulence, severe icing, and widespread dust or sandstorms that reduce visibility to less than three miles.

Simplified Directional Facility/SDF—A NAVAID used for nonprecision instrument approaches. The final approach course is similar to that of an ILS localizer except that the SDF course may be offset from the runway, generally not more than 3 degrees, and the course may be wider than the localizer, resulting in a lower degree of accuracy.

Special VFR conditions—Weather conditions in a control zone which are less than basic VFR and in which some aircraft are permitted flight under Visual Flight Rules.

Special VFR operations—Aircraft operating in accordance with clearances within control zones in weather conditions less than the basic VFR weather minima. Such operations must be requested by the pilot and approved by ATC.

speed adjustment—An ATC procedure used to request pilots to adjust aircraft speed to a specific value for the purpose of providing desired spacing. Speed adjustments are always expressed as indicated airspeed and pilots are expected to maintain a speed of plus or minus 10 knots of the specified speed.

squawk (mode, code, function)—Activate specific modes/codes/functions on the aircraft transponder, e.g., "Squawk Three/Alpha, Two one zero five, Low."

Standard Instrument Departure/SID—A preplanned instrument flight rule (IFR) air traffic control departure procedure printed for pilot use in graphic and/or textual form. SIDs provide transition from the terminal to the appropriate en route structure.

Standard Terminal Arrival Route/STAR—A preplanned instrument flight rule (IFR) air traffic control arrival route published for pilot use in graphic and/or textual form. STARs provide transition from the en route structure to a fix or point from which an approach can be made.

stand by—Means the controller or pilot must pause for a few seconds, usually to attend to other duties of a higher priority. Also means to "wait" as in "stand by for clearance." If a delay is lengthy, the caller should reestablish contact.

stop altitude squawk—Used by ATC to inform an aircraft to turnoff the automatic altitude reporting feature of its transponder. It is issued when the verbally reported altitude varies 300 feet or more from the automatic altitude report.

stop squawk (mode or code)—Used by ATC to tell the pilot to turn specified functions of the aircraft transponder off.

straight-in approach-IFR—An instrument approach wherein final approach is begun without first having executed a procedure turn. Not necessarily completed with a straight-in landing or made to straight-in landing minimums.

straight-in approach-VFR—Entry into the traffic pattern by interception of the extended runway centerline (final approach course) without executing any other portion of the traffic pattern.

straight-in landing—A landing made on a runway aligned within 30° of the final approach course following completion of an instrument approach.

surveillance approach—An instrument approach wherein the air traffic controller issues instructions, for pilot compliance, based on aircraft position in relation to the final approach course (azimuth), and the distance (range) from the end of the runway as displayed on the controllers radar scope. The controller will provide recommended altitudes on final approach *if* requested by the pilot.

taxi into position and hold—Used by ATC to inform a pilot to taxi onto the departure runway in takeoff position and hold. It is not authorization for takeoff. It is used when takeoff clearance cannot immediately be issued because of traffic or other reasons.

terminal area—A general term used to describe airspace in which approach control service or airport traffic control service is provided.

terminal radar program—A national program instituted to extend the terminal radar services provided IFR aircraft to VFR aircraft. Pilot participation in the program is urged but is not mandatory. The progressive stages of the program are referred to as Stage I, Stage II and Stage III.

1. *Stage I/Radar Advisory Service for VFR Aircraft*—Provides traffic information and limited vectoring to VFR aircraft on a workload permitting basis.

2. *Stage II/Radar Advisory and Sequencing for VFR aircraft*—Provides, in addition to Stage I service, vectoring and sequencing on a full-time basis to arriving VFR aircraft. The purpose is to

adjust the flow of arriving IFR and VFR aircraft into the traffic pattern in a safe and orderly manner and to provide traffic advisory to departing VFR aircraft.

3. *Stage III/Radar Sequencing and Separation Service for VFR Aircraft*—Provides, in addition to State II services, separation between all participating aircraft. The purpose is to provide separation between all participating VFR aircraft and all IFR aircraft operating within the airspace defined as a Terminal Radar Service Area (TRSA), or Terminal Control Area (TCA).

Terminal Radar Service Area/TRSA—Airspace surrounding designated airports wherein ATC provides radar vectoring, sequencing and separation on a full-time basis for all IFR and participating VFR aircraft. Service provided in a TRSA is called Stage III Service. Graphics depicting TRSA layout and communications frequencies are shown in the AIM. Pilot participation is urged but is not mandatory.

tetrahedron—A device normally located on uncontrolled airports and used as a landing direction indicator. The small end of a tetrahedron points in the direction of landing. At controlled airports, the tetrahedron, if installed, should be disregarded because tower instructions supersede the indicator.

that is correct—The understanding you have is right.

threshold—The beginning of that portion of the runway usable for landing.

thunderstorm intensity levels—Existing radar systems cannot detect turbulence per se. However, because there is a direct correction between thunderstorm precipitation density and storm intensity, the NWS is able to measure the strength of radar weather echoes for the purpose of categorizing the strength of the storm:

Level 1 (WEAK) and Level 2 (MODERATE). Moderate to severe turbulence possible.

Level 3 (STRONG). Severe turbulence possible, lightning.

Level 4 (VERY STRONG). Severe turbulence likely, lightning.

Level 5 (INTENSE). Severe turbulence, lightning, organized wind gusts. Hail likely.

Level 6 (EXTREME). Severe turbulence, large hail, lightning, extensive wind gusts and turbulence.

time group—Four digits representing the hour and minutes from the 24-hour clock. Time group without time zone indicators are understood to be GMT (Greenwich Mean Time); e.g., "0205." A time zone indicator is used to indicate local time, e.g., "0205M." The end and beginning of the day are shown by "2400" and "0000" respectively.

time in service—With respect to maintenance time records, means the time from the moment an aircraft leaves the surface of the earth until it touches it at the next point of landing.

touchdown—
>1. The point at which an aircraft first makes contact with the landing surface.
>2. Concerning a precision radar approach (PAR), it is the point where the glide path intercepts the landing surface.

touchdown zone—The first 3000 feet of the runway beginning at the threshold.

Touchdown Zone Elevation/TDZE—The highest elevation in the first 3000 feet of the landing surface. TDZE is indicated on the instrument approach procedure chart when straight-in landing minimums are authorized.

tower/airport traffic control tower—A terminal facility which through the use of air/ground communications, visual signaling, and other devices, provides ATC services to airborne aircraft operating in the vicinity of an airport and to aircraft operating on the movement area.

track—The actual flight path of an aircraft over the surface of the earth.

tower en route control service/tower to tower—The control of IFR en route traffic within delegated airspace between two or more adjacent approach control facilities. This service is designed to expedite traffic and reduce control and pilot communication requirements.

traffic advisories—Advisories issued to alert a pilot to other known or observed IFR/VFR air traffic which may be in such proximity to his aircraft's position or intended route of flight to warrant his attention. Such advisories may be based on:
>1. Visual observation from a control tower,
>2. Observation of radar identified and nonidentified aircraft targets on an ARTCC/Approach Control radar scope, or,
>3. Verbal reports from pilots or other facilities.

Controllers use the word "traffic" followed by additional information, if known, to provide such advisories, e.g., "Traffic, 2 o'clock, one zero miles, southbound, fast moving, altitude readout seven thousand five hundred." Traffic advisory service will be provided to the extent possible depending on higher priority duties of the controller or other limitations, e.g., radar limitations, volume of traffic, frequency congestion or controller workload. Radar/nonradar traffic advisories do not relieve the pilot of his responsibility for continual vigilance to see and avoid other aircraft. IFR and VFR aircraft are cautioned that there are many times when the controller is not able to give traffic advisories concerning all traffic in the aircraft's proximity; in other words, when a pilot requests or is receiving traffic advisories, he should not assume that all traffic will be issued.

traffic alert, advise you turn right/left heading (degrees) and/or climb/descent to (altitude) immediately—See *safety advisory*

traffic in sight—Used by pilots to inform a controller that previously-issued traffic is in sight.

traffic pattern—The traffic flow that is prescribed for aircraft landing at, taxiing on, or taking off from an airport. The components of a typical traffic pattern are upwind leg, crosswind leg, downwind leg, base leg and final approach.

> 1. *Upwind Leg*—A flight path parallel to the landing runway in the direction of landing.
>
> 2. *Crosswind Leg*—A flight path at right angles to the landing runway off its upwind end.
>
> 3. *Downwind Leg*—A flight path parallel to the landing runway in the direction opposite to landing. The downwind leg normally extends between the crosswind leg and the base leg.
>
> 4. *Base Leg*—A flight path at right angles to the landing runway off its approach end. The base leg normally extends from the downwind leg to the intersection of the extended runway centerline.
>
> 5. *Final Approach*—A flight path in the direction of landing along the extended runway centerline. The final approach normally extends from the base leg to the runway. An aircraft making a straight-in approach VFR is also considered to be on final approach.

Transcribed Weather Broadcast/TWEB—A continuous recording of metrological and aeronautical information that is broadcast on L/MF and VOR facilities for pilots.

transmitting in the blind—A transmission from one station to other stations in circumstances where two-way communication cannot be established but where it is believed that the called stations may be able to receive the transmission.

transponder—The airborne radar beacon receiver/transmitter portion of the Air Traffic Control Radar Beacon System (ATCRBS) which automatically receives radio signals from interrogators on the ground, and selectively replies with a specific reply pulse or pulse group only to those interrogations being received on the mode to which it is set to respond.

T-VOR/Terminal-Very High Frequency Omnidirectional Range Station—A very high frequency terminal omnirange station located on or near an airport and used as an approach aid.

ultrahigh frequency/UHF—The frequency band between 300 and 3000 MHz. The band of radio frequencies used for military air/ground voice communications. In some instances this may go as low as 225 MHz and still be referred to as UHF.

unable—Indicates inability to comply with a specific instruction, request, or clearance.

uncontrolled airspace—That portion of the airspace that has *not* been designated as continental control area, control area, control zone, terminal control area, or transition area.

UNICOM—A non-government air/ground radio communication facility which may provide airport advisory service at certain airports.

vector—A heading issued to an aircraft to provide navigational guidance by radar.

verify—Request confirmation of information; e.g., "verify assigned altitude."

very high frequency/VHF—The frequency band between 30 and 300 MHz. Portions of this band, 108 to 118 MHz, are used for certain NAVAIDS; 118 to 136 MHz are used for civil air/ground voice communications. Other frequencies in this band are used for purposes not related to air traffic control.

VFR not recommended—An advisory provided by a Flight Service Station to a pilot during a preflight or inflight weather briefing that flight under visual flight rules is not recommended. To be given when the current and/or forecasted weather conditions are at or below VFR minimums. It does not abrogate the pilot's authority to make his own decision.

VFR on top—An IFR clearance used in lieu of a specified altitude assignment upon pilot's request which authorizes the aircraft to be flown in VFR weather conditions at an appropriate VFR altitude/flight level which is not below the minimum IFR altitude.

VFR over the top—The operation of an aircraft over-the-top under VFR when it is not being operated on an IFR flight plan.

visibility—The ability, as determined by atmospheric conditions and expressed in units of distance, to see and identify *prominent* unlighted objects by day and *prominent* lighted objects by night. Visibility is reported as statute miles, hundreds of feet or meters.

 1. Flight Visibility—The average forward horizontal distance, from the cockpit of an aircraft in flight, at which prominent unlighted objects may be seen and identified by day and prominent lighted objects may be seen and identified by night.

 2. Ground Visibility—Prevailing horizontal visibility near the earth's surface as reported by the United States National Weather Service or an accredited observer.

 3. Prevailing Visibility—The greatest horizontal visibility equaled or exceeded throughout *at least half* the horizon circle which *need not necessarily be continuous.*

 4. Runway Visibility Value/RVV—The visibility determined for a particular runway by a transmissometer. A meter provides a continuous indication of the visibility (reported in miles or fraction of miles) for the runway. RVV is used in lieu of prevailing visibility in determining minimums for a particular runway.

5. *Runway Visual Range/RVR*—An instrumentally derived value, based on standard calibrations, that represents the horizontal distance a pilot will see down the runway from the approach end; it is based on the sighting of either high intensity runway lights or on the visual contrast of other targets whichever yields the greater visual range. RVR, in contrast to prevailing or runway visibility, is based on what a pilot in a moving aircraft should see looking down the runway. RVR is horizontal visual range, not slant visual range. It is based on the measurement of a transmissometer made near the touchdown point of the instrument runway and is reported in hundreds of feet.

    a. *Touchdown RVR*—The RVR visibility readout values obtained from RVR equipment serving the runway touchdown zone.

    b. *MID RVR*—The RVR readout values obtained from RVR equipment located midfield of the runway.

    c. *Rollout RVR*—The RVR readout values obtained from RVR equipment located nearest the rollout end of the runway.

visual approach—An approach wherein an aircraft on an IFR flight plan, operating in VFR conditions under the control of an air traffic control facility and having an air traffic control authorization, may proceed to the airport of destination in VFR conditions.

Visual Flight Rules/VFR—Rules that govern the procedures for conducting flight under visual conditions. The term "VFR" is also used in the United States to indicate weather conditions that are equal to or greater than minimum VFR requirements. In addition, it is used by pilots and controllers to indicate type of flight plan.

visual separation—A means employed by ATC to separate IFR aircraft in terminal areas. There are two ways to effect this separation:

    1. The tower controller sees the aircraft involved and issues instructions, as necessary, to ensure that the aircraft avoid each other.

    2. A pilot sees the other aircraft involved and upon instructions from the controller provides his own separation by maneuvering his aircraft as necessary to avoid it. This may involve following another aircraft or keeping it in sight until it is no longer a factor.

VORTAC/VHF Omnidirectional Range/Tactical Air Navigation—A navigation aid providing VOR azimuth, TACAN azimuth, and TACAN distance measuring equipment (DME) at one site. (See *VOR, Distance Measuring Equipment, Navigational Aid*)

vortices/wingtip vortices—Circular patterns of air created by the movement of an airfoil through the air when generating lift. As an airfoil moves through the atmosphere in sustained flight, an area of high pressure is created beneath it and an area of low pressure is created above it. The air flowing from the high pressure area to the low pressure area around and about the tips of the airfoil tends to roll up into two rapidly rotating vortices, conical in shape. These vortices are the most predominant parts of aircraft wake turbulence and their rotational force is dependent upon the wing loading, gross weight, and speed of the generating aircraft. The vortices from medium to heavy aircraft can be of extremely high velocity and hazardous to smaller aircraft.

VOR/Very High Frequency Omnidirectional Range Station—A ground-based electronic navigation aid transmitting very high frequency navigation signals, 360 degrees in azimuth, oriented from magnetic north. Used as the basis for navigation in the national airspace system. The VOR periodically identifies itself by morse code and may have an additional voice identification feature. Voice features may be used by ATC or FSS for transmitting instructions/information to pilots.

VOT/VOR test signal—A ground facility which emits a test signal to check VOR receiver accuracy. The system is limited to ground use only.

wake turbulence—Phenomena resulting from the passage of an aircraft through the atmosphere. The term includes vortices, thrust stream turbulence, jet blast, jet wash, propeller wash and rotor wash, both on the ground and in the air.

Warning Area—Specified international airspace within which there may exist activities constituting a potential danger to aircraft. Warning areas are depicted on aeronautical charts.

waypoint—See *Area Navigation*.

wilco—I have received your message, understand it, and will comply with it.

wind shear—A change in wind speed and/or wind direction in a short distance, resulting in a tearing or shearing effect. It can exist in a horizontal or vertical direction and occasionally in both.

words twice—

1. As a request: "Communication is difficult. Please say every phrase twice."

2. As information: "Since communications are difficult, every phrase in this message will be spoken twice."

COMMON VFR-IFR AVIATION ABBREVIATIONS AND SYMBOLS

ALS—Approach light system.
ASR—Airport surveillance radar.
ATC—Air traffic control.

CAS—Calibrated airspeed.

DH—Decision height.
DME—Distance measuring equipment compatible with TACAN.

FM—Fan marker.

GS—Glide slope.

HIRL—High-intensity runway light system.

IAS—Indicated airspeed.
IFR—Instrument flight rules.
ILS—Instrument landing system.
IM—ILS inner marker.
INT—Intersection.

LDA—Localizer-type directional aid.
LFR—Low frequency radio range.
LMM—Compass locator at middle marker.
LOC—ILS localizer.
LOM—Compass locator at outer marker.

MAA—Maximum authorized IFR altitude.
MCA—Minimum crossing altitude.
MDA—Minimum descent altitude.
MEA—Minimum en route IFR altitude.
MM—ILS middle marker.
MOCA—Minimum obstruction clearance altitude.
MRA—Minimum reception altitude.
MSL—Mean sea level.

NDB (ADF)—Nondirectional beacon (automatic direction finder).
NOPT—No procedure turn required.

OM—ILS outer marker.

PAR—Precision approach radar.

RAIL—Runway alignment indicator light system.
RBN—Radio beacon.
RCLM—Runway centerline marking.
RCLS—Runway centerline light system.
REIL—Runway end identification lights.
RVR—Runway visual range as measured in the touchdown zone area.

SALS—Short approach light system.
SSALS—Simplified short approach light system.

TAS—True airspeed.
TDZL—Touchdown zone lights.
TVOR—Very high frequency terminal omnirange station.

V_A—Design maneuvering speed.
V_F—Design flap speed.
V_{FE}—Maximum flap extended speed.
V_{LE}—Maximum landing gear extended speed.
V_{LO}—Maximum landing gear operating speed.
V_{MC}—Minimum control speed with the critical engine inoperative.
V_{MO}—Means maximum operating limit speed.
V_{NE}—Never-exceed speed.
V_{SO}—The stalling speed or the minimum steady flight speed in the landing configuration.
V_{S1}—The stalling speed or the minimum steady flight speed obtained in a specified configuration.
V_X—Speed for best angle of climb.
V_Y—Speed for best rate of climb.
V_{SSE}—Minimum safe single engine speed. Unless this speed is maintained a twin engine aircraft cannot reach V_{YSE} without giving up altitude.

A BASIC IFR CLEARANCE SHORTHAND

In taking an IFR clearance, it is common that some form of shorthand be used to keep up with the clearance as transmitted by the controller. At this time, there is no universally agreed-upon shorthand form; the matter is left up to the individual pilot. The following shorthand symbols and abbreviations are therefore simply suggested forms. However, they do represent typical contractions in use today.

To become proficient in IFR shorthand, the key is practice. Copying clearances by listening to a recording of practice clearances or to ATC clearance delivery will save dollars by minimizing cockpit time. At today's IFR training price of almost one dollar per minute, a "say again" is expensive indeed. Learn and practice the symbols and abbreviations at home—a few evenings are all that is required. The mark of a professional is "clearance correct" to a readback, after having been given a clearance only once.

BASIC IFR CLEARANCE SYMBOLS

$\text{\AA}$ = Airport
$\odot$ = VOR
$\multimap\!\!\odot$ = Enter Control Zone
$\multimap\!\!\ominus\!\!\rightarrow$ = Thru Control Zone
$\ominus\!\!\rightarrow$ = Leave Control Zone

C/50 or	↑	= Climb to*
M/50 or	→	= Cruise*
D/50 or	↓	= Descend to*
LT or	⌐	= Left Turn
RT or	⌐	= Right Turn
	∩	= Reverse Course
CM/50 or	⌐	= Climb to and Maintain*
	=	= While in Civil Airways
	⊣	= Enter Airways
	╫	= Cross Airways
	┼	= Intercept
┼ or	△	= Intersection
	5̄0̄	= Remain at or below*
	5̲0̲	= Remain at*
	5̲0̲	= Remain at or above*
PHXAC or	⒜⒫	= Contact Approach Control**
PHXC or	⒫ⓗⓧ	= Contact Center**
PHXDC or	⒟ⓒ	= Contact Departure Control**
	Ɛ	= And
	°	= Deg. Magnetic
	/	= To, over or word separation

BASIC IFR CLEARANCE ABBREVIATIONS

A	= After
AD	= After Departure
AL	= After Leaving
ALT	= Altitude
AP	= After Passing
AEH	= After Entering Holding
ADV	= Advise
APP	= Approaching
B	= Before
BC	= Back Course
BP	= Before Passing
BR	= Before Reaching
C	= ATC Clears
C	= Climb
CRS	= Course
CT	= Contact (or advise)
CVA	= Clearance Void After

---

* The figure 50 is used simply as an example of an altitude; in this case, 5000 feet above sea level.

**The notation PHX is used simply as an example of a three-letter station identifier.

D	=	Direct
D	=	Descend
DLA	=	Delay
DLAI	=	Delay Indefinite
DEP	=	Depart
DR	=	Direct
E	=	East
EAC	=	Expect Approach Clearance
EAT	=	Expect Approach Time
EFC	=	Expect Further Clearance
EHA	=	Expect Higher Altitude
ETA	=	Est. Time of Arrival
ETE	=	Est. Time En Route
ETD	=	Est. Time of Departure
F	=	Final
FAF	=	Final Approach Fix
FM	=	Fan Marker
G	=	GMT
H	=	Hold
HDG	=	Heading
I	=	Immediately
IB	=	Inbound
L	=	Localizer
L	=	Left
LS	=	Left Side
LOM	=	Locator Outer Marker
M	=	Maintain
m	=	Minutes
NM	=	Nautical Miles
N	=	North
NDE	=	No Delay Expected
NLT	=	No Later Than
OB	=	Outbound
OM	=	Outer Marker
OVR	=	Over
OR	=	If not possible
P or FPR	=	Flight Plan Route
PT	=	Passing Through
R	=	Right
R or RAD	=	Omni Radial
R	=	Report
RO or R/PHX	=	Report Over**
RP	=	Report Passing
RX	=	Report Crossing
RCRS	=	Report on Course

**The notation PHX is used simply as an example of a three-letter station identifier.

RL	=	Report Leaving
RR	=	Report Reaching
RSPT	=	Report Start Procedure Turn
RC	=	Revise Course
RW	=	Runway
RH	=	Maintain Runway Heading
RV	=	Radar Vector
RV	=	Radar Verification
RS	=	Right Side
S	=	South
SI	=	Straight-In Approach
STBY	=	Standby
TFC	=	Traffic
T/O	=	Take-off
tl	=	Until
TWR	=	Tower
UFA	=	Until Further Advised
V	=	Victor Airway
V or VTR	=	Vector
V	=	Verify
V	=	Void
VFR/H	=	Example "on top" clearance, i.e., VFR on top of haze layer
W	=	West
W	=	Wind
WHN	=	When
X	=	Cross
Z	=	Zulu (time)

RELATED IFR ABBREVIATIONS

MAA	=	Max Authorized Alt
MEA	=	Min En route Alt
MCA	=	Min Crossing Alt
MHA	=	Min Holding Alt
MRA	=	Min Radio Alt
MDA	=	Min Descent Alt
MAP	=	Missed Approach Point
HAT	=	Height Above Touchdown
HAA	=	Height Above Airport
DH	=	Decision Height

TYPICAL IFR CLEARANCE FORMAT

1. Clearance Limit

2. En route Instructions
3. Altitude, Special Instructions
4. Holding
5. Take-off Instructions
6. Departure Control Frequency
7. Squawk Code

Index

Index